Synthesis Lectures on Renewable Energy Technologies

The series, Synthesis Lectures on Renewable Energy Technologies publishes concise books, focused on technologies that harness energy from naturally occurring sources, such as sunlight, wind, water, geothermal heat, and biofuels from organic materials. These renewable energy technologies play a crucial role in transitioning away from fossil fuels, helping to mitigate the effects of climate change, and promoting a sustainable energy supply.

Richard A. Dunlap

Renewable Energy Storage

Mechanical and Thermal Methods

Springer

Richard A. Dunlap
Department of Physics
and Atmospheric Science
Dalhousie University
Halifax, NS, Canada

ISSN 2690-5000 ISSN 2690-5019 (electronic)
Synthesis Lectures on Renewable Energy Technologies
ISBN 978-3-031-88944-8 ISBN 978-3-031-88942-4 (eBook)
https://doi.org/10.1007/978-3-031-88942-4

This Springer imprint is published by the registered company Springer Nature Switzerland AG
The registered company address is: Gewerbestrasse 11, 6330 Cham, Switzerland

Preface

At present approximately 80% of our energy worldwide comes from the combustion of fossil fuels. This approach to energy is not sustainable because of the limited fossil fuel resources available. As well, the need to change to non-fossil fuel energy sources is accentuated by the adverse environmental effects of continued fossil fuel use. Most notable of the environmental consequences of fossil fuel use is global climate change. Although the transition to renewable carbon-free energy sources is essential, it is not easy. A significant aspect of the use of renewable energy sources is the need for energy storage. Most renewable energy sources are neither constant in time, nor are they readily portable. These two features are a requirement for much of our energy use. Specifically, a reliable supply of heat and electricity is needed for residential, as well as commercial and industrial needs, and a portable source of energy is essential for most transportation applications.

The present book considers some of the important technologies for energy storage that utilize mechanical methods and thermal methods to store energy. Chapter 1 considers pumped hydroelectric energy storage and Chap. 2 considers compressed air energy storage. The use of gravitational potential of solid masses and flywheels to store energy is presented in Chap. 3. Chapter 4 reviews the use of sensible heat to store thermal energy. These concepts are expanded upon in Chap. 5, where solar ponds, which act as both solar collectors and thermal energy storage devices, are considered. Finally, Chap. 6 discusses the use of the latent heat of materials as an energy storage mechanism.

Halifax, Canada Richard A. Dunlap

Contents

1 Pumped Hydroelectric Energy Storage 1
1.1 Introduction 1
1.2 Conventional Pumped Hydroelectric Storage 1
1.3 Pump-Back Hydroelectric Storage 6
1.4 Seawater-Based Pumped Hydroelectric Storage 8
1.5 Sub-surface Pumped Hydroelectric Storage 12
1.6 Underwater Reservoirs 13
1.7 World Use of Pumped Hydroelectric Storage 15
References 20

2 Compressed Air Energy Storage 23
2.1 Introduction 23
2.2 Physics of Compressed Air Energy Storage 23
2.3 Locations for Grid Scale Compressed Air Energy Storage 26
2.4 Use of Compressed Air Grid Storage 30
2.5 Other Applications of Compressed Air Energy Storage 32
2.6 Combined Pumped Hydroelectric-Compressed Air Energy Storage 36
References 39

3 Energy Storage Using Solid Masses 43
3.1 Introduction 43
3.2 Gravitational Potential of Solid Masses 43
3.2.1 Advanced Rail Energy Storage 44
3.2.2 Gravity Power Module 45
3.2.3 GraviStore 46
3.2.4 Energy Vault 47
3.3 Flywheels 49
3.3.1 The Physics of Flywheels 50
3.3.2 Flywheel Design Criteria 54

3.3.3 Applications of Flywheels for Energy Storage 57
References 63

4 Sensible Heat Energy Storage 67
4.1 Introduction 67
4.2 Basics of Sensible Heat Energy Storage 67
4.3 Sensible Heat Energy Storage Materials 68
4.4 Residential Heat Storage 69
4.5 District Heating Systems 74
4.5.1 Reykjavik District Heating System 75
4.5.2 Drake Landing Solar Community 77
4.5.3 Brædstrup Fjernvarme 80
4.6 Grid Integrated Systems 84
References 88

5 Solar Ponds 91
5.1 Introduction 91
5.2 Basic Principles of Solar Ponds 91
5.3 Applications of Solar Ponds 93
5.3.1 Solar Pond Heat Removal 93
5.3.2 Industrial Heat 94
5.3.3 Space Heating 96
5.3.4 Desalination 96
5.3.5 Electricity Generation 98
5.4 History of Solar Ponds 99
5.4.1 Beit HaArava Solar Pond, West Bank 99
5.4.2 El Paso Solar Pond, Texas 100
5.4.3 Bhuj Solar Pond, Gujarat, India 100
5.4.4 Pyramid Hill, Victoria, Australia 101
5.4.5 Granada Solar Pond, Spain 101
References 102

6 Latent Heat and Thermochemical Energy Storage 105
6.1 Introduction 105
6.2 Basics of Latent Heat Energy Storage 105
6.3 Solid–Liquid Systems 108
6.3.1 Latent Heat Energy Storage in Ice 108
6.3.2 Other Solid–Liquid Phase Change Materials 109
6.4 Latent Heat Energy Storage in Liquid Air 111
6.5 Thermochemical Energy Storage 112
6.5.1 Reactions 115
6.5.2 Adsorption 116
References 120

About the Author

Richard A. Dunlap received a B.S. in Physics (Worcester Polytechnic Institute 1974), an A.M. in Physics (Dartmouth College 1976) and a Ph.D. in Physics (Clark University 1981). Since receiving his Ph.D., he has been on the Faculty at Dalhousie University where he currently holds an appointment as Research Professor in the Department of Physics and Atmospheric Science. Prof. Dunlap has more than 300 refereed research publications in fields that include critical phenomena, magnetic materials, amorphous alloys, quasicrystals, hydrogen storage and advanced battery materials. His previously published books include *Experimental Physics: Modern Methods* (Oxford 1988), *The Golden Ratio and Fibonacci Numbers* (World Scientific 1997), *Novel Microstructures for Solids* (IOP/Morgan & Claypool 2018), *Particle Physics* (IOP/Morgan & Claypool 2018), *Sustainable Energy—2nd ed.* (Cengage 2019), *Energy from Nuclear Fusion* (IOP Publishing 2021), *Transportation Technologies for a Sustainable Future* (IOP Publishing 2023), *Lasers and their Application in the Cooling and Trapping of Atoms—2nd ed.* (IOP Publishing 2023), *An Introduction to the Physics of Nuclei and Particles—2nd ed.* (IOP Publishing 2023), *The Mössbauer Effect—2nd ed.* (IOP Publishing 2024), *Generation IV Nuclear Reactors: Design, Operation and Prospects for Future Energy Production* (IOP Publishing 2024) and *Renewable Energy—Requirements and Sources* (Springer Nature 2025).

Pumped Hydroelectric Energy Storage 1

1.1 Introduction

Mechanical energy storage methods include several diverse techniques. These are used primarily for the grid-scale storage of electrical energy but applications to transportation have also been considered. The most common grid storage technology is pumped hydroelectric storage. Here electricity is used to pump water to a higher gravitational potential in order to store energy. This energy can be recovered by allowing the water to run back to a lower elevation through a turbine. The present chapter reviews the physics of pumped hydroelectric energy storage and discusses the development and growth of this technology.

1.2 Conventional Pumped Hydroelectric Storage

In a conventional pumped hydroelectric storage facility, water flows between an upper reservoir and a lower water supply (reservoir, river, lake or ocean) where the upper reservoir is supplied only by water pumped from the lower reservoir. The reservoirs may be natural or artificial. The most common configuration uses an artificial upper reservoir and a natural water source (e.g., river) as the lower reservoir. If the lower reservoir is connected to a natural water source, then the system is referred to as an open-loop system. If neither reservoir has a source of water other than that which is cycled through the pumps/turbines, then the system is referred to as a closed-loop system. These types of systems are illustrated in Fig. 1.1.

Overall storage efficiency is limited by the motor/pump efficiency, turbine/generator efficiency and water loss in the upper reservoir due to evaporation. Net efficiencies are typically in the 70–80% range (compared to conventional hydroelectric generating facilities which operate in the 85–90% efficiency range). Since the storage and recovery of

R. A. Dunlap, *Renewable Energy Storage*, Synthesis Lectures on Renewable Energy Technologies, https://doi.org/10.1007/978-3-031-88942-4_1

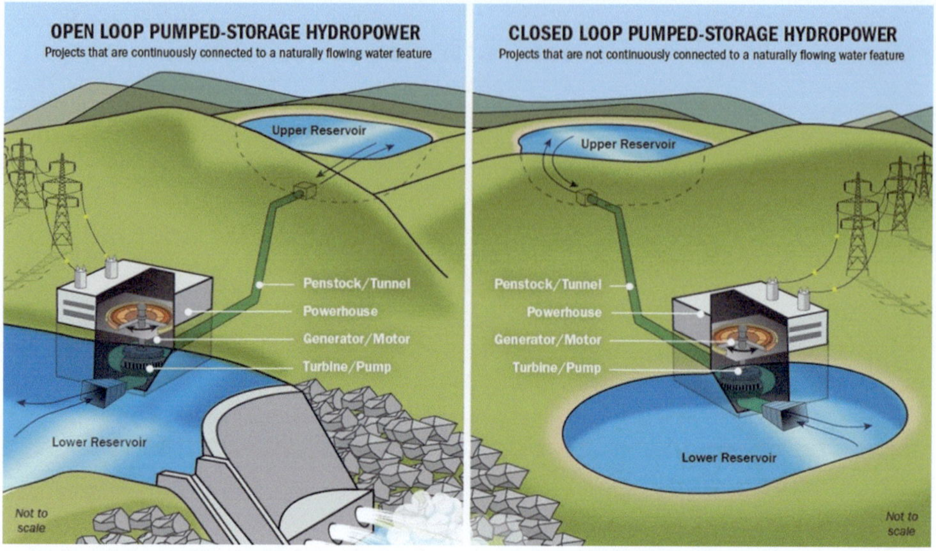

Fig. 1.1 Differences between open-loop (left) and closed-loop (right) pumped hydroelectric storage systems. U. S. Department of Energy (ND) Public domain

electrical energy requires both pumping (to store the energy) and generation (to recover the energy) the overall efficiency is the result of these combined processes.

Pumped hydroelectric storage is a commonly used method of topping up grid electricity during times of higher demand. It is a common method of load leveling or peak shaving, that is storing energy during periods of low demand and recovering this energy during periods of high demand (see Dunlap 2025). Pumped hydroelectric storage is used to store electricity which has been generated by any method, not just hydroelectricity or other renewable methods. It is a convenient means of grid storage as facilities have substantial generating capacity (power), as well as considerable total energy storage capacity. In addition, this power can be brought on-line quickly to satisfy demand.

The typical design of a pumped hydroelectric facility is illustrated in Fig. 1.2. Water is pumped from the lower reservoir through a penstock to the upper reservoir using the motor/pump in order to store energy. Electrical energy is recovered when water from the upper reservoir flows through the penstock to the turbine/generator near the lower reservoir. If the average head between the upper reservoir and the generator is h, then the total energy available from the gravitational potential of the water in the upper reservoir is

$$E = mgh = \rho ghV. \tag{1.1}$$

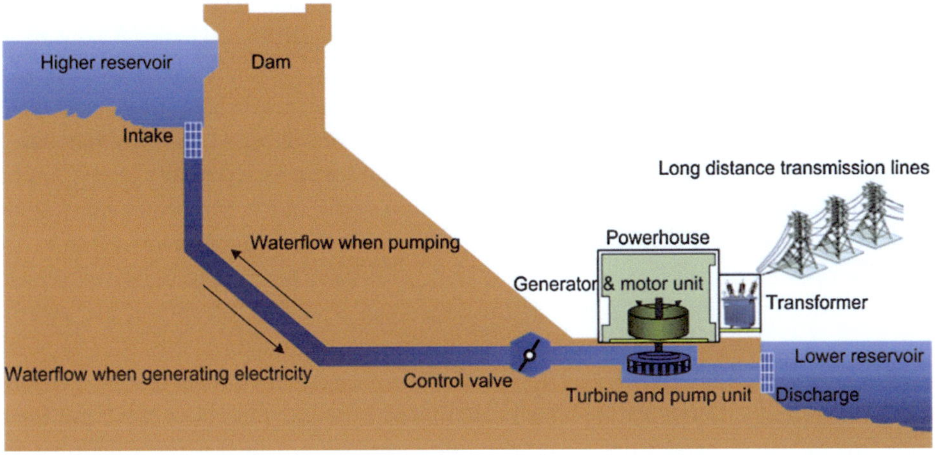

Fig. 1.2 Diagram of a typical pumped hydroelectric storage facility. Figure 4 from Luo et al. (2015) Copyright (2015) the Authors. CC BY 3.0. https://creativecommons.org/licenses/by/3.0/

Here m is the total mass of water in the upper reservoir, g is the gravitational acceleration, ρ is the density of the water (kg/m^3) and V is the total volume of the upper reservoir (m^3). When h is in meters then the energy is in Joules. Including the net system efficiency, η, this may be written as

$$E = \eta \rho g h V. \tag{1.2}$$

The power generated by the turbine, in Watts, is

$$P = \frac{dE}{dt} = \eta \rho g h \frac{dV}{dt} = \eta \rho g h \varphi, \tag{1.3}$$

where φ is the flow rate in m^3/s. If the total penstock cross sectional area is A, in m^2, then the flow rate is given in terms of the water velocity in the penstock, v, as

$$\varphi = vA. \tag{1.4}$$

Thus, Eq. (1.3) may be written as

$$P = \eta \rho g h v A. \tag{1.5}$$

The above equations can also be used to estimate the total time, t, the facility can provide maximum power. Since $P = E/t$ then $t = E/P$ or from Eqs. (1.2) and (1.5),

$$t = \frac{V}{vA}, \tag{1.6}$$

where, using the SI units above, the time is given in seconds. The power is frequently expressed in MW and the energy capacity in MWh where MWh = (Joules)/(3.6×10^9).

An analysis of Eqs. (1.2), (1.5) and (1.6) provides the relevant characteristics of a pumped hydroelectric storage facility, that is, the total energy storage capacity, the maximum power available and the duration of maximum power. In Eq. (1.2), η is a function of the motor/pump and turbine/generator design, while ρ and g are constants ($\rho = 1000$ kg/m^3 for fresh water and 1025 kg/m^3 for sea water, $g = 9.8$ m/s^2). The total energy stored is, therefore, a linear function of the average head and the upper reservoir volume. The maximum power available, as given in Eq. (1.5), is a linear function of the water velocity in the penstock and the total penstock cross sectional area. The velocity of the water in the penstock is, for practical purposes, limited to about 6 m/s and penstocks are typically set at a grade of about 100% (i.e. 45° from the horizontal). This means that the penstock area is the primary design feature which determines the maximum power available. Equation (1.6) follows directly from a consideration of the volume of water and the flow rate and shows that the ratio of the upper reservoir volume to the penstock area will determine the duration of power available.

Figure 1.3 shows a typical artificial reservoir that is used as the upper water source for a pumped hydroelectric storage system, i.e., the Porąbka-Żar Power Plant in Poland. The typical internal structure of an artificial reservoir used for pumped hydroelectric storage is shown in Fig. 1.4. This figure shows the reservoir of the 300 MW Mount Gilboa Pumped Storage in Israel during construction. The generators of the Ludington Pumped Hydroelectric facility in Michigan are shown during construction in Fig. 1.5. This facility was constructed during the period of 1969 to 1973, and, at a capacity of 1872 MW, it was the world's largest pumped hydroelectric storage facility at the time. It is now the fifth largest pumped hydroelectric facility. The tops of the six 362 MW Francis pump/turbines are seen in the figure. Figure 1.6 shows the pump/turbine facility at the Raccoon Mountain Pumped Hydroelectric Facility in Tennessee. The facility consists of four 413 MW generators for a total generating capacity of 1652 MW.

In many cases, water for pumped hydroelectric facilities is fed through penstocks that are located underground, as in the case of the Ludington and Raccoon Mountain facilities described above. Figure 1.7 shows an example of above ground penstocks connecting the upper and lower reservoirs of a pumped hydroelectric facility, in this case the 1507 MW Castaic Pumped-Storage Plant. This facility uses Pyramid Lake, which has been formed by a dam across the Piru Creek as the upper reservoir. The lower reservoir is Elderberry Forebay, which was formed by damming off a portion of Castaic Lake.

There are basically three different configurations that are used for pumped hydroelectric storage, depending on the number of turbines, pumps, motors and generators that are used. These are the *binary set*, *ternary set* and *quaternary set* configurations, as described below.

Fig. 1.3 Porąbka-Żar Power Plant in Międzybrodzie Bialskie, Poland. This is a 500 MW pumped hydroelectric facility which became operational in 1979. The lower reservoir is the Soła River. Ongrys (2008) CC BY-SA 3.0. https://creativecommons.org/licenses/by-sa/3.0/deed.en

- The binary set configuration utilizes one combined pump/turbine and one combined motor/generator unit on the same shaft. This is by far the most common design, mainly because it is the most cost effective. However, it is not the most efficient, as the pump/turbine design is inevitably a compromise between the optimal design for these two purposes.
- The ternary set configuration uses a pump, motor/generator and turbine mounted on the same shaft. The use of a separate pump and turbine allows for the design of both of these components to be optimized.
- Quaternary set configuration uses a motor-pump assembly mounted on one shaft and a turbine-generator assembly mounted on another shaft. The two units operate independently and are, in general, located in separate powerhouses.

Fig. 1.4 The 300 MW Mount Gilboa Pumped Storage under construction. The facility is located 60 km to the east of Haifa, Israel. Note the wind farm in the distance. McKaby (2016) CC BY-SA 4.0. https://creativecommons.org/licenses/by-sa/4.0/deed.en

1.3 Pump-Back Hydroelectric Storage

Pumped hydroelectric storage that is incorporated into a conventional hydroelectric facility is referred to as a pump-back facility. In this case, an upper reservoir, formed by constructing a dam across a river, can be augmented in times of low demand by pumping water into the reservoir from the river below the dam. An example of a pump-back storage facility utilizing above ground penstocks is illustrated in Fig. 1.8. This is the Tumut 3 generating station in New South Wales, Australia. This station has an average head of 151 m and six 300 MW turbines for a total capacity of 1800 MW. Three of the six turbines can be operated as pumps for pumped hydroelectric storage. The upper Talbingo Reservoir was formed by constructing a dam on the Tumut River.

The Grand Coulee Dam on the Columbia River in Washington State is a notable example of a hydroelectric generating station with a pump-back facility. The Grand Coulee Dam was initially designed in 1935 as a means of irrigation control, but subsequently, in 1941, became operational as a hydroelectric facility. Over the years the conventional hydroelectric generating capacity has been increased and is currently at 6495 MW supplied by 27 Francis turbine-generators. Between 1973 and 1984, six motor-pump/turbine-generator units were installed for a total pump-back capacity of 314 MW. This pump-back facility is named the John W. Keys III Pump-Generating Plant and is shown

Fig. 1.5 The Ludington Pumped Hydroelectric facility under construction. Sequeira (2015) Public domain

in Fig. 1.9. A pipe, referred to as the tailrace, connects the pumphouse to the lower reservoir. The pump-back plant cycles water between the Franklin D. Roosevelt Lake on the down-stream side of the dam and Banks Lake on the up-stream side of the dam with a head of 85 m.

Fig. 1.6 Generator facility at the Raccoon Mountain Pumped Hydroelectric Storage Facility in Tennessee. Construction of the Raccoon Mountain station began in 1970 and was completed in 1978. The generating facility has a net capacity of 1616 MW. Rankin (2016) Public domain. The appearance of U.S. Department of Defense (DoD) visual information does not imply or constitute DoD endorsement

1.4 Seawater-Based Pumped Hydroelectric Storage

In principle, the source of water for a pumped hydroelectric storage plant can be the ocean rather than a river, lake or reservoir. To date, one pumped hydroelectric storage facility that utilizes seawater has been constructed (Fujihara et al. 1998). The Okinawa Yanbaru Seawater Pumped Storage Power Station in Kunigami, Japan was a 30 MW facility that operated between 1999 and 2016. Specifications for this facility are summarized in Table 1.1. Details of the intake/outlet in the Philippine Sea are shown in Fig. 1.10. The intake/outlet is protected by a breakwater constructed of dolosse (concrete tetrapods). While the Okinawa station has demonstrated the possibility of using seawater in a pumped hydroelectric storage plant, it also emphasized some of the difficulties of using seawater in a hydroelectric facility. These difficulties include:

Fig. 1.7 Castaic Pumped-Storage Plant located in Los Angeles County, CA showing the above ground penstocks. Sirbatch (2011) CC BY-SA 3.0. https://creativecommons.org/licenses/by-sa/3.0/deed.en

- The corrosive properties of seawater (compared with fresh water). This problem required certain steel components that are in contact with the seawater to be replaced with stainless steel components. The use of cathodic protection on some of the equipment was also found to be important.
- The fouling of certain components with marine organisms, specifically barnacles. It was found that barnacles do not adhere to surfaces for water flow rates greater than about 5 m/s. For portions of the system where flow rates are substantially less than this value, surfaces were coated with hydrophobic paint, which inhibits barnacle adhesion.
- Infiltration of seawater into the land from the upper reservoir.
- Water intake/outlet during variable sea condition.

Following from the experience gained from the Okinawa station with the use of seawater in a hydroelectric facility, several other pumped hydroelectric storage facilities have been considered, although none is operational. Two plans for possible future seawater based hydroelectric energy storage are described below.

Espejo de Tarapacá—Espejo de Tarapacá is a planned seawater pumped hydroelectric storage facility in the town of Pintados, Chile approximately 75 km southeast of Iquique (Valhalla ND). It is to be located on a 600 m high coastal cliff in the Atacama Desert and would have a capacity of 300 MW. The upper reservoir would use a natural concavity in the desert lined with an impermeable membrane. The Pacific Ocean would serve as the

Fig. 1.8 Tumut 3 hydroelectric generating station in New South Wales, Australia showing penstocks. Cmh (2006) CC BY-SA 3.0. https://creativecommons.org/licenses/by-sa/3.0/deed.en

lower reservoir. Espejo de Tarapacá would be constructed in conjunction with Cielos de Tarapacá, a 600 MW solar photovoltaic facility in the Atacama Desert.

Dundrum, Northern Ireland—A proposed seawater pumped hydroelectric storage facility in Dundrum, Northern Ireland (McLean and Kearney 2014) would take a very different approach than the projects mentioned above. This would be a 100 MW low head high flow rate facility created by constructing a dam across the mouth of a small ~ 6 km^2 sheltered bay. The bay would form the upper reservoir, and the ocean would act as the lower reservoir. The maximum head would be about 6 m and the maximum discharge would be 2066 m^3/s. This type of facility shows similarities with barrage-based tidal energy stations, such as The Annapolis Royal Generating Station (Tethys Engineering ND) in Nova Scotia. This facility was operational from 1984 to 2019 and had a capacity of 20 MW. It utilized the natural tidal range of up to 7.5 m as the head and flow rates of up to 400 m^3/s.

Fig. 1.9 John W. Keys III Pump-Generating Plant on the Columbia River in Washington State. U.S. Department of the Interior, Bureau of Reclamation (2012). CC BY-SA 2.0. https://creativecommons.org/licenses/by-sa/2.0/

Table 1.1 Specifications of the Okinawa Yanbaru Seawater Pumped Storage Power Station

Quantity	Value
Maximum output	31.4 MW
Net head	141 m
Maximum discharge	26 m^3/s
Upper reservoir depth	22.8 m
Upper reservoir volume	5.9×10^5 m^3
Penstock length	314 m
Penstock diameter	24 m
Tailrace length	205 m
Tailrace diameter	27 m

Data adapted from Fujihara et al. (1998) and Hino and Lejeune (2012)

Fig. 1.10 Seawater intake/outlet on the Philippine Sea of the Okinawa Yanbaru Seawater Pumped Storage Power Station. gpzagogo (2010) Public domain

1.5 Sub-surface Pumped Hydroelectric Storage

In addition to the designs described above, the necessary head between two water sources can be obtained by using a surface level upper reservoir and an underground lower reservoir. The general concept of the sub-surface pumped hydroelectric storage facility is illustrated in Fig. 1.11. While no facilities of this type have been constructed, a number of possible future projects have been proposed. A couple of proposed facilities are discussed below.

Muuga Harbour, Estonia—The planned energy storage facility at Muuga Harbour, Estonia (about 13 km northeast of the capital city of Tallinn) makes use of both the concept of seawater pumped hydroelectric storage and sub-surface pumped hydroelectric storage (Energiasalv Pakri OÜ 2010). This project utilizes seawater, obtained from Muuga Harbour as the upper reservoir and a cavern excavated in granite bedrock as the lower reservoir. From the intake, seawater travels about 1.7 km through the 7-m dimeter penstock to an underground reservoir at a depth of 530–565 m below the surface. Energy is stored by pumping water out of the underground reservoir and into the ocean. Energy

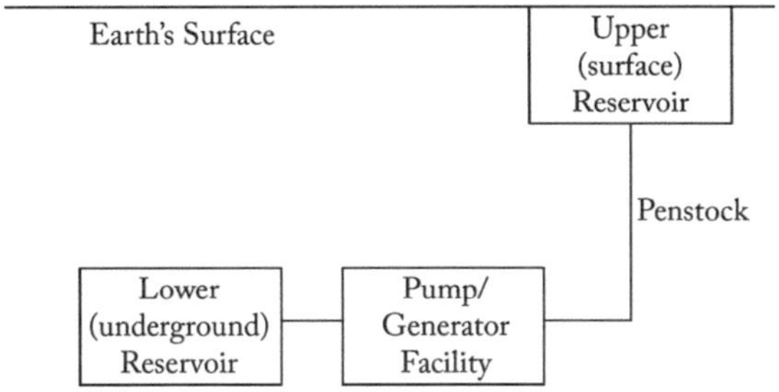

Fig. 1.11 General design of an underground (sub-surface) pumped hydroelectric storage facility

is recovered by allowing seawater to flow from the ocean to the underground reservoir. The planned turbine facility will consist of one 100 MW and two 175 MW reversible vertical-shaft Francis pump/turbines and one 50 MW vertical-shaft Francis turbine. The total generating capacity will be 500 MW and nominal output can be maintained for a period of about 12 h.

Bendigo, Victoria, Australia—While the sub-surface approach may, in general, appear to be more complex than the traditional elevated upper reservoir design, the possibility of utilizing abandoned mines as the lower reservoir may actually make this design more cost effective. As well, the depth of many possible mine sites would provide a head that is considerably greater than that associated with traditional pumped hydroelectric stations. Bendigo, Victoria, Australia has numerous abandoned gold mines from the late nineteenth century, with depths up to around 1400 m. It has been suggested (ARUP 2024) that many of these shafts would be suitable for subsurface pumped hydroelectric storage facilities.

1.6 Underwater Reservoirs

The use of an underwater chamber as a storage reservoir is an interesting variation on the conventional concept of pumped hydroelectric storage that has been considered in recent years. A basic diagram of this concept, known as ORES (Ocean Renewable Energy Storage), is shown in Fig. 1.12. The chamber, in this case a spherical concrete chamber, is located on the ocean floor at a great depth. Energy is stored when the water is pumped out of the storage volume and electricity is generated by the turbine when water is allowed back into the chamber. It is important to realize that the buoyancy of the chamber, when it is evacuated, must be counteracted in order to keep the chamber on the ocean floor. Thus, sufficient ballast must be incorporated into the design of the device for this purpose.

Following from the mathematical description of conventional pumped hydroelectric energy, as given above, the energy capacity of the ORES storage chamber is

$$E = \eta \rho g d V \quad (1.7)$$

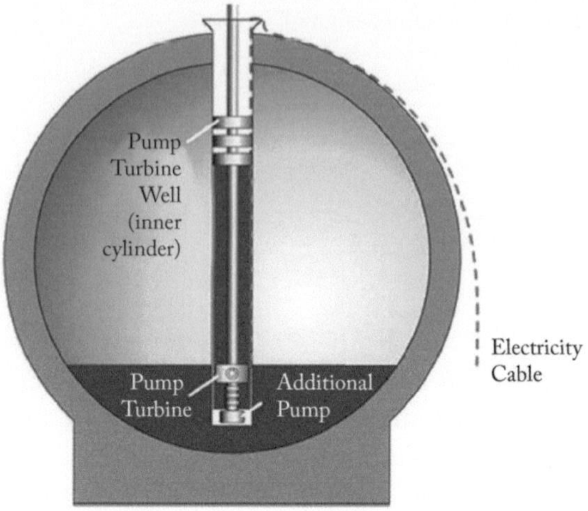

Fig. 1.12 Diagram of the ORES concept. Figure 1 Reprinted from Hahn et al. (2017) Copyright (2017) with permission of Elsevier

where, as above, η is the net system efficiency, ρ is the density, in this case the density of sea water, 1025 kg/m^3 and g is the gravitational acceleration. In Eq. (1.7), d is the depth of the storage chamber in the ocean and V is the inner volume of the chamber.

From a practical standpoint, the volume of the storage chamber will be substantially less than the volume of the upper reservoir in a conventional pumped hydroelectric storage plant. On the other hand, the water depth may be greater than the typical head in the conventional system. Overall, however, one might expect the total storage capacity to be substantially less than that for existing pumped hydroelectric facilities.

Some typical design parameters of an ORES unit are given in Table 1.2. The table shows that the unit as described has a total energy storage capacity of about 15 MWh. By comparison a typical conventional pumped hydroelectric storage facility, e.g. Raccoon Mountain, as described above, has a storage capacity of about 35 GWh. The ORES concept, however, may have some advantages over other approaches to energy storage. They may be constructed as energy storage farms consisting of multiple units, rather than a single unit, thereby increasing capacity proportionately. Since they are located underwater, they may be a convenient means of storing electricity from offshore wind farms. The proximity to the generating devices has clear advantages and the combination of offshore wind turbines with underwater storage may minimize terrestrial environmental impact. This type of storage unit is in the early experimental stages but may play a role in future renewable energy development.

Table 1.2 Typical design parameters and calculated energy storage capacity for an ORES unit

Parameter	Symbol	Value	Units
Depth	d	500	m
Sphere diameter	–	30	m
Sphere volume	V	14,137	m^3
Net efficiency	η	0.75	–
Energy storage capacity	E	14.8	MWh
Power generated	P	5	MW
Duration of power	t	2.96	h

1.7 World Use of Pumped Hydroelectric Storage

Pumped hydroelectric storage is commonly used for grid connected systems in order to even out the load during times of varying demand. This process is "load leveling" or "peak shaving" depending on the details of the use of stored energy and has been described in some detail by Dunlap (2024). An example of how stored energy can be used for load leveling during the varying daily electricity demand is illustrated in Fig. 1.13. The figure shows low demand during the night when excess generated electricity can be stored. Between about 8:00 am and noon the demand grows and remains large until early evening when it decreases again. During this period energy that is stored in pumped hydroelectric storage is used to top up energy generated by fossil fuel or renewable sources. Since pumped hydroelectric storage is less than 100% efficient, the total electricity that needs to be generated is greater than without load leveling, but the maximum capacity (power) that is required is less, see the net electricity shown in the figure.

While load leveling generally refers to the situation shown in Fig. 1.13 and is used to reduce the maximum power that needs to be produced, peak shaving refers to the situation where the net load on the generating system remains reasonably constant throughout the day. This is illustrated in Fig. 1.14.

Pumped hydroelectric storage is also an effective means of utilizing nuclear energy. Current nuclear power reactor designs are generally not capable of adjusting their output rapidly in response to load fluctuations. Pumped storage is, therefore, an effective way of utilizing nuclear as a contribution to base load grid capacity. With the continued development of renewable energy sources, pumped hydroelectric storage has also provided an effective method of integrating intermittent sources such as solar and wind with fossil fuel-based grid capacity.

While several of the technologies mentioned in the present chapter are still in the early development stages, traditional pumped hydroelectric storage has made substantial contributions to grid storage and, at present, represents the most highly developed energy storage method for this purpose. While infrastructure costs of pumped hydroelectric storage can be quite high, maintenance is generally minimal, and the lifetime of facilities is

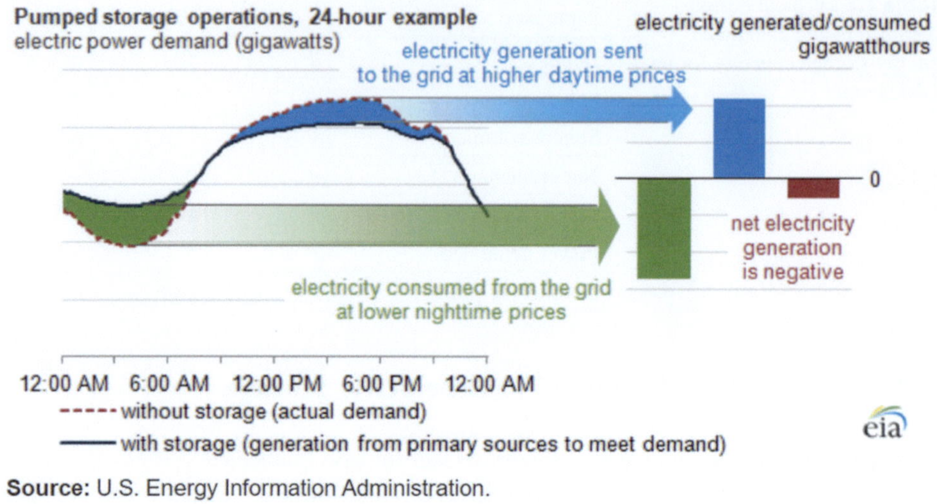

Fig. 1.13 Example of the daily use of pumped hydroelectric storage showing nighttime storage of excess electricity and daily use of stored energy. U. S. Energy Information Administration (2013) Public domain

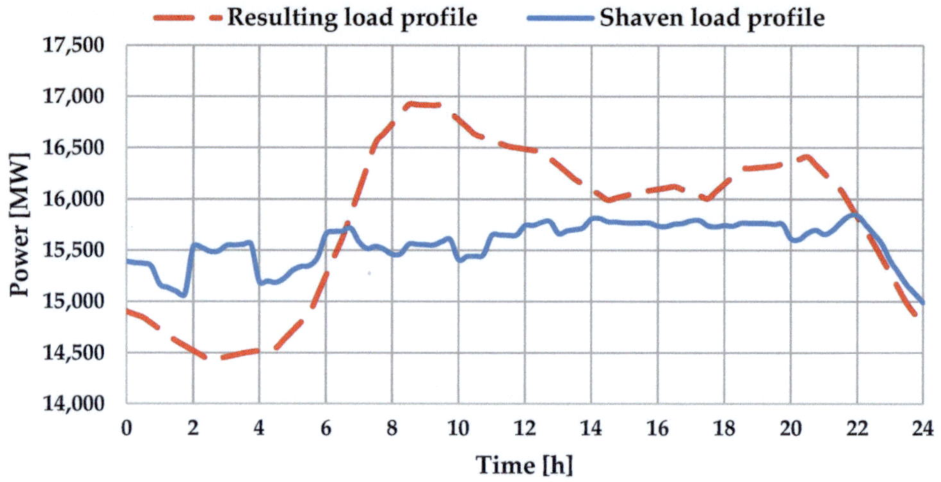

Fig. 1.14 Typical example of electricity load with peak shaving using pumped hydroelectric storage. Figure 6 from Šćekić et al. (2020) Copyright (2020) by the Authors CC BY 4.0. https://creativecommons.org/licenses/by/4.0/

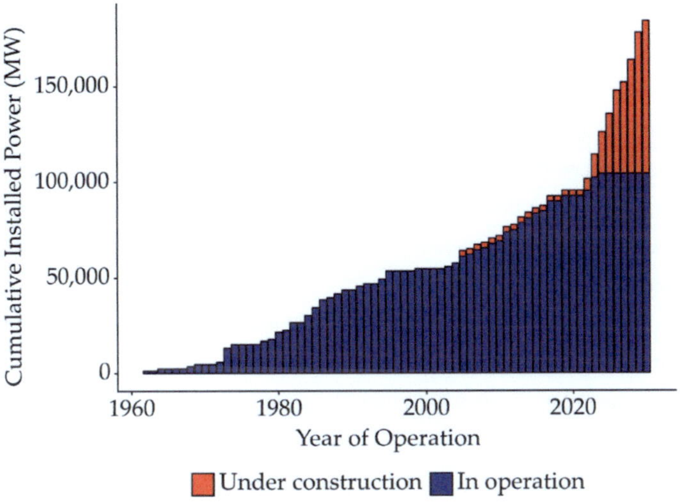

Fig. 1.15 Pumped hydroelectric storage capacity since 1960 showing operational capacity and capacity under construction. Figure 3 from Nikolaos et al. (2023) Copyright (2023) by the Authors CC BY 4.0. https://creativecommons.org/licenses/by/4.0/

quite long. There has been considerable growth in pumped hydroelectric storage capacity in recent years. This is illustrated in Fig. 1.15.

The world's largest pumped hydroelectric storage facilities, based on maximum generating capacity, are summarized in Table 1.3. As the table shows the largest facilities have generating capacities of more than 2 GW, compared to coal-fired and nuclear generating stations, which are typically around 1 GW.

It is clear that the majority of large pumped hydroelectric facilities are in China. It is also obvious that these have all opened within the past 20 years or so. In fact, the growth of pumped hydroelectric storage capacity, as shown in Fig. 1.15, can be illustrated on a basis of country, as shown in Fig. 1.16. Here it is clear that recent pumped hydroelectric storage capacity development has been, almost exclusively, in China. This is also clear from the breakdown of in-operation and under-construction capacity by country as summarized in Table 1.4.

The United States has a long history of pumped hydroelectric storage where the first commercial use of this approach dates from 1930. In this year, the Connecticut Electric and Power Company opened a storage facility near New Milford, Connecticut. Water was pumped from the Housatonic River to a storage reservoir with a head of 70 m. There was considerable development in the 1970s and 1980s (see Table 1.3) with the opening of the Ludington Pumped Storage Power Plant in Michigan (1973) and the Bath County Pumped Storage Station in Virginia (1985). The Bath County Pumped Storage Station (see Fig. 1.17) was the largest pumped hydroelectric storage facility when it opened and

Table 1.3 The largest, by generating capacity, pumped hydroelectric storage stations in the world

Station	Country	Year[a]	Capacity (MW)
Fengning Pumped Storage Power Station	China	2019	3600
Bath County Pumped Storage Station	United States	1985	3003
Huizhou Pumped Storage Power Station	China	2007	2448
Guangdong Pumped Storage Power Station	China	1994	2400
Meizhou Pumped Storage Power Station	China	2023	2400
Ludington Pumped Storage Power Plant	United States	1973	2172
Changlongshan Pumped Storage Power Station	China	2021	2100
Okutataragi Pumped Storage Power Station	Japan	1974	1932
Tianhuangping Pumped Storage Power Station	China	2004	1836
Tumut-3	Australia	1973	1800

[a]Time of first operation. In many cases stations added capacity over time, so, they were often not at full capacity at the time of first operation

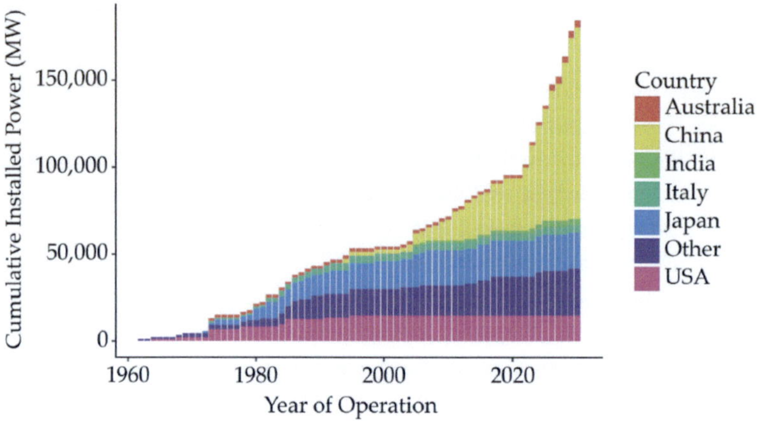

Fig. 1.16 Growth of pumped hydroelectric storage capacity by country. Figure 4 from Nikolaos et al. (2023) Copyright (2023) by the Authors CC BY 4.0. https://creativecommons.org/licenses/by/4.0/

remained so for more than thirty years when it surpassed in capacity by the Fengning Pumped Storage Power Station in Hebei Province, China in 2021. As Fig. 1.16 shows, there has been virtually no further development of pumped hydroelectric storage facilities in the United States since the 1980s.

The distribution of pumped hydroelectric storage facilities in the United States is illustrated on the map in Fig. 1.18. It is seen that the majority of facilities are located along the East Coast and in California. Facilities near the East Coast typically make use of the

Table 1.4 Capacity of operational and under construction pumped hydroelectric storage by country

Country	Operational capacity (MW)	Capacity under construction (MW)
China	40,648	69,550
Japan	15,307	2820
United States	13,731	0
Italy	4200	0
Australia	1800	2250
Ukraine	2531	900
Taiwan	2608	0
United Kingdom	2500	0
Egypt	0	2400
South Africa	2332	0

Data adapted from Nikolaos et al. (2023) and references therein

Fig. 1.17 Bath County Pumped Storage Station in Bath County, Virginia. Z22 (2023) CC BY-SA 4.0. https://creativecommons.org/licenses/by-sa/4.0/deed.en

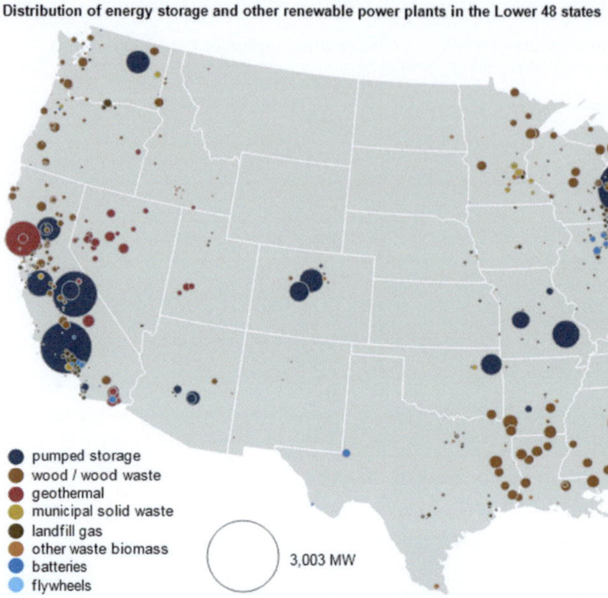

Fig. 1.18 Map of pumped hydroelectric storage facilities (and other renewable facilities) in the United States showing the generating capacity of each plant by the size of the circle. U. S. Energy Information Administration (2017) Public domain

advantageous geography offered by the Appalachian Mountains. In fact, the Bath County facility lies near the Eastern Continental Divide on the border between Virginia and West Virginia.

References

ARUP (2024) Transforming an old mine into renewable-enabling energy storage. https://www.arup.com/projects/bendigo-underground-pumped-storage/

Cmh (2006) Tumut3 generating station. https://commons.wikimedia.org/wiki/File:Tumut3GeneratingStation.jpg

Dunlap RA (2025) Renewable energy—requirements and sources. Springer Nature, Cham. https://doi.org/10.1007/978-3-031-77185-9

Energiasalv Pakri OÜ (2010) Brief description of the Muuga seawater-pumped hydro accumulation power plant. Project ENE 1001. http://energiasalv.ee/wp-content/uploads/2012/07/Muuga_HAJ_17_02_2010_ENG.pdf

Fujihara T, Imano H, Oshima K (1998) Development of pump turbine for seawater pumped storage power plant. Hitachi Rev 47:199–202. http://coopeoliennes.free.fr/fichiers/Okinawa.pdf

gpzagogo (2010) The seawater intake/outlet of the Okinawa Yanbaru Seawater pumped storage power station. https://commons.wikimedia.org/wiki/File:Seawater_intake-outlet_of_Yanbaru_Power_Station.jpg

Hahn H, Hau D, Dick C, Puchta M (2017) Techno-economic assessment of a subsea energy storage technology for power balancing services. Energy 133:122–125. https://doi.org/10.1016/j.energy.2017.05.116

Hino T, Lejeune A (2012) Pumped storage hydropower developments. In: Sayigh A (ed) Comprehensive renewable energy, vol 6, pp 405–434. https://doi.org/10.1016/B978-0-08-087872-0.00616-8

Luo X, Wang J, Dooner M, Clarke J (2015) Overview of current development in electrical energy storage technologies and the application potential in power system operation. Appl Energy 137:511–536. https://doi.org/10.1016/j.apenergy.2014.09.081

McKaby (2016) Ma'ale Gilboa Pumped-storage hydroelectricity under construction. https://commons.wikimedia.org/wiki/File:Ma%27ale_Gilboa_Pumped-storage_hydroelectricity_under_constraction.jpg

McLean E, Kearney D (2014) An evaluation of seawater pumped hydro storage for regulating the export of renewable energy to the national grid. Energy Proc 46:152–160. https://doi.org/10.1016/j.egypro.2014.01.168

Nikolaos PC, Marios F, Dimitris K (2023) A review of pumped hydro storage systems. Energies 16:4516. https://doi.org/10.3390/en16114516

Ongrys (2008) Zbiornik elektrowni szczytowo-pompowej na górze Żar widziany z kabiny szybowca. https://commons.wikimedia.org/wiki/File:Zar_zbiornik.jpg

Rankin M (2016) Silver Jackets team tours the Raccoon mountain pumped-storage plant (Image 1 of 6). https://www.dvidshub.net/image/2937779/silver-jackets-team-tours-raccoon-mountain-pumped-storage-plant

Šćekić L, Mujović S, Radulović V (2020) Pumped hydroelectric energy storage as a facilitator of renewable energy in liberalized electricity market. Energies 2020:6076. https://doi.org/10.3390/en13226076

Sequeira P (2015) Generating station of consumer power plant in Ludington. Muskegon, Michigan. https://commons.wikimedia.org/wiki/File:GENERATING_STATION_OF_CONSUMER_POWER_PLANT_IN_LUDINGTON_-_NARA_-_547129_--_color_graded.tif?page=1

Sirbatch (2011) Castaic Power Plant front. https://commons.wikimedia.org/wiki/File:Castaic_Power_Plant_Front.jpg

Tethys Engineering (ND) Annapolis Tidal Station. https://tethys.pnnl.gov/project-sites/annapolis-tidal-station

U. S. Department of Energy (ND) Pumped Storage Hydropower. https://www.energy.gov/eere/water/pumped-storage-hydropower

U.S. Department of the Interior, Bureau of Reclamation (2012) John W. Keys III Pump-generating plant. https://www.flickr.com/photos/usbr/6925644278

U. S. Energy Information Administration (2013) Pumped storage provides grid reliability even with net generation loss. https://www.eia.gov/todayinenergy/detail.php?id=11991

U. S. Energy Information Administration (2017) Energy storage and renewables beyond wind, hydro, solar make up 4% of U.S. power capacity. https://www.eia.gov/todayinenergy/detail.php?id=31372

Valhalla (ND) Espejo de Tarapacá. http://valhalla.cl/espejo-de-tarapaca/

Z22 (2023) Bath County Pumped Storage Station. https://commons.wikimedia.org/wiki/File:Bath_County_Pumped_Storage_Station.jpg

Compressed Air Energy Storage

2

2.1 Introduction

In addition to the pumped hydroelectric energy storage techniques described in the previous chapter, grid scale storage may also be accomplished by using electricity to pressurize a gas. The energy that is stored may be recovered by allowing the gas to flow through a turbine during decompression. Similar techniques can be used to store energy on a smaller scale, and these have been considered for applications such as vehicle propulsion. It is essential to look in detail at the thermodynamics of the pressurization and de-pressurization in order to understand the functioning of a compressed air energy storage system. This chapter begins by looking at the basic physics of compressed air energy storage. The application of this technique to grid storage and smaller scale systems is considered. Finally, an energy storage system that combines compressed air energy storage with pumped hydroelectric energy storage is described.

2.2 Physics of Compressed Air Energy Storage

The basic concept of compressed air energy storage (CAES) is quite simple. Electricity is used to operate a motor-pump to compress air in a confined volume. The air is then expended through a turbine, which turns a generator to recover the stored electricity. However, in practice the process is not so simple and not so easy. In order to understand how compressed air energy storage actually works, we need to look in some detail at the thermodynamics of compressing a gas.

We consider the adiabatic compression of a gas. In an adiabatic process the system is thermally isolated from its surroundings and no heat is transferred. The first law of thermodynamics gives the change in total internal energy of a gas, ΔU, in terms of the

R. A. Dunlap, *Renewable Energy Storage*, Synthesis Lectures on Renewable Energy Technologies, https://doi.org/10.1007/978-3-031-88942-4_2

heat supplied to the system, ΔQ, and the work done by the system, W, as

$$\Delta U = \Delta Q - W, \tag{2.1}$$

where the work done by the system may be written as

$$W = P\Delta V. \tag{2.2}$$

For an adiabatic process, $\Delta Q = 0$, which gives

$$\Delta U = -P\Delta V. \tag{2.3}$$

For an ideal gas the internal energy is given by

$$U = c_v nT, \tag{2.4}$$

where c_v is the molar heat capacity, n is the number of moles and T is the temperature. Combining Eqs. (2.3) and (2.4) gives

$$c_v n\Delta T = -P\Delta V. \tag{2.5}$$

For compression, $\Delta V < 0$ and the change in temperature will be $\Delta T > 0$, i.e. heating during compression. For expansion $\Delta V > 0$, so $\Delta T < 0$ and the gas will cool. In terms of compressed air energy storage, the work performed by the compressor both compresses the gas and heats it. When energy is recovered by expanding the gas (to turn a turbine) the gas cools and all of the energy input into the system is returned (except for losses due to friction etc. in the machinery). The temperature of the gas after compression, T_a, relative to the temperature of the gas before compression, T_b, may be related to the pressures before, P_b, and after, P_a, compression as

$$T_a = T_b\left(\frac{P_a}{P_b}\right)^{\chi}, \tag{2.6}$$

where the exponent χ is related to the ratio of specific heats at constant pressure, c_P, and at constant volume, c_V, as

$$\chi = \left(\frac{c_P}{c_V} - \frac{c_V}{c_P}\right). \tag{2.7}$$

For dry air $\chi \approx 0.29$. In practice, this means that the temperature of the compressed gas can be quite high, and this may cause practical problems.

In practice, compression of a gas can be adiabatic, diabatic or isothermal. As noted above, adiabatic compression refers to the case where there is no heat transfer between the system and the environment, so that the system temperature is defined by Eq. (2.6). In principle, a system could be designed so that the heat produced during compression was transferred to a thermal storage medium (e.g. stone, oil, salt, etc., see Chaps. 4 and 6). In this case the stored heat can be returned to the gas during the expansion phase to increase the efficiency of energy recovery.

Diabatic compression refers to the situation where the heat associated with compression is partially removed from the gas by means of heat exchangers (referred to as intercoolers). The compressed gas remains at a temperature that is less than for the adiabatic case. In practice, this means that the gas must be heated prior to expansion. This is typically done using a natural gas- fired heater.

Finally, isothermal compression refers to the situation where all heat generated by the compression is transferred to the environment, thereby keeping the temperature of the compressed gas the same as before the compression.

A simple way of estimating the maximum energy storage capacity in a compressed gas, is to consider the isothermal case. The gas is described by the ideal gas law,

$$PV = nRT, \tag{2.8}$$

where n is the number of moles of gas and R is the ideal gas constant. The work done on the system to compress the gas from an initial pressure and volume, P_i and V_i, to a final pressure and volume, P_f and V_f, is given by

$$E = \int_{V_f}^{V_i} PdV = \int_{V_f}^{V_i} \frac{nRT}{V} dV = P_f V_f \cdot \ln\left(\frac{P_f}{P_i}\right), \tag{2.9}$$

and is equal to the theoretical energy stored. A simple calculation shows that the energy stored in a 1 m^3 volume for an initial pressure of 100 kPa (1 atmosphere) and a final pressure of (say) 10 MPa (100 atmospheres) is

$$E = (10\,\text{MPa}) \times \ln\left(10^7/10^5\right) = 46\,\text{MJ}, \tag{2.10}$$

or about 13 kWh. This provides a rough estimate of the maximum energy capacity in a compressed air energy storage facility.

Figure 2.1 shows a schematic of a compressed air energy storage facility. In this case, the gas is compressed diabatically and, as noted above, must be heated during expansion. During expansion the gas is mixed with natural gas and combusted. The expanding gas turns the natural gas turbine to drive the generator and produce electricity.

Several different approaches can be taken to this process. Figure 2.2 shows a simple diabatic compressed air energy storage system where the heat of compression is dissipated as waste heat. During energy recovery, the compressed air is expanded, mixed with natural gas and combusted to drive a single stage turbine. Figure 2.3 shows an adiabatic system where the heat of compression is not dissipated but is stored in the compressed air. In this case no additional heat is needed during energy recovery. Figure 2.4 shows

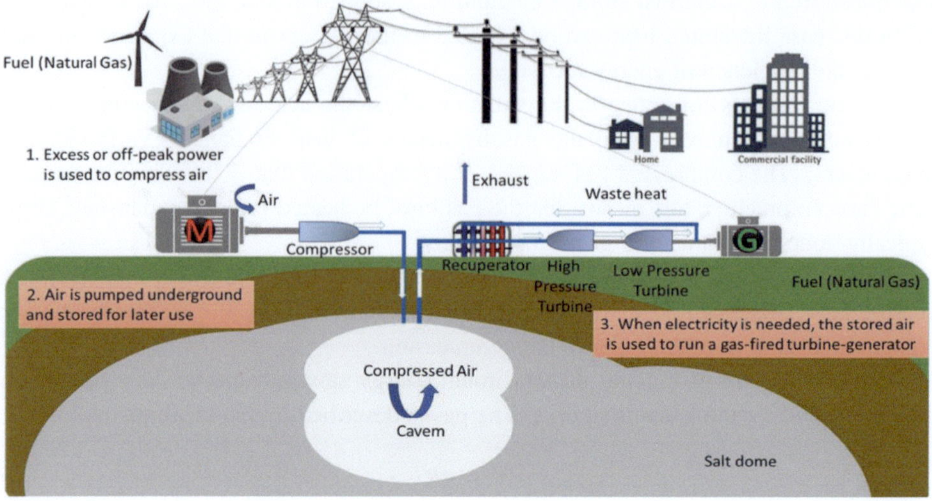

Fig. 2.1 Schematic of a compressed air energy storage facility. Figure 3 from Rabi et al. (2023) Copyright (2023) by the Authors CC BY 4.0. https://creativecommons.org/licenses/by/4.0/

an isothermal system where the compressed air is maintained at a constant temperature during energy storage and recovery. A system that uses a two-stage turbine arrangement where the exhaust from the first high pressure turbine is combined with additional fuel and combusted to drive the second low pressure turbine is illustrated in Fig. 2.5. Excess heat from the exhaust of the low pressure turbine is recovered with a recuperator.

2.3 Locations for Grid Scale Compressed Air Energy Storage

In the present section we consider the requirements for the implementation of a grid-scale compressed air energy storage system. It is clear from the above discussion that grid-scale compressed air energy storage requires the use of considerable pressures. More importantly, however, is the need for an appropriately large volume that will withstand the required pressures. There are several possibilities for suitable locations. These fall into two basic categories: underground locations and underwater locations. The most convenient underground locations for compressed air energy storage are geological formations associated with salt domes. A salt dome is a natural formation where a column of salt intrudes upward into an overlaying layer of sedimentary rock. Figure 2.6 shows a diagram of a salt dome. Formations often have associated oil deposits as shown in the figure. A cavern appropriate for compressed air energy storage can be created by solution-mining the salt dome. This process involves dissolving the salt by flushing the deposit with water. Caverns prepared in this manner have long been used for the storage of gases and liquids

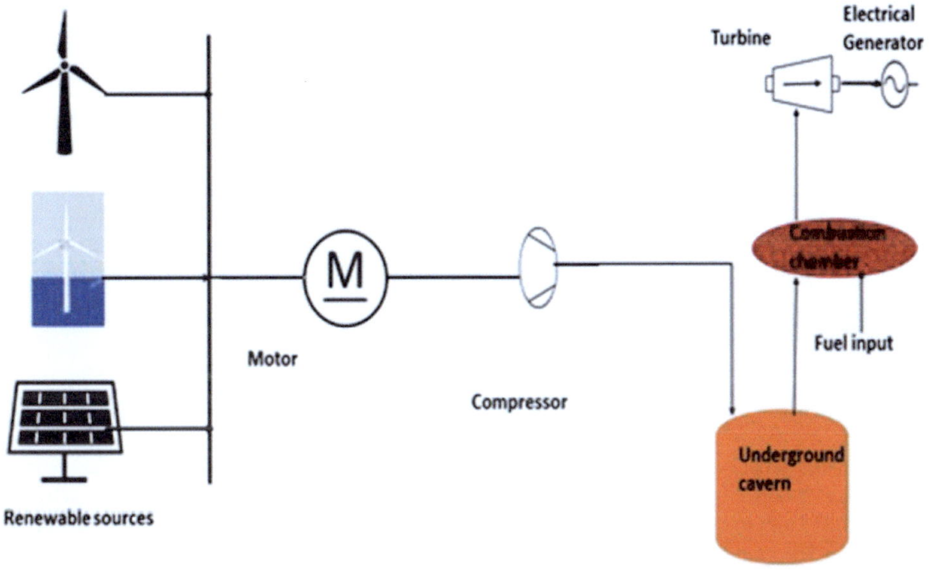

Fig. 2.2 Diabatic compressed air energy storage system where the heat of compression is dissipated as waste heat. Figure 1(a) from Venkataramani et al. (2018) Copyright (2018) The Authors. Reprinted with permission of Springer CC BY 4.0. https://creativecommons.org/licenses/by/4.0/

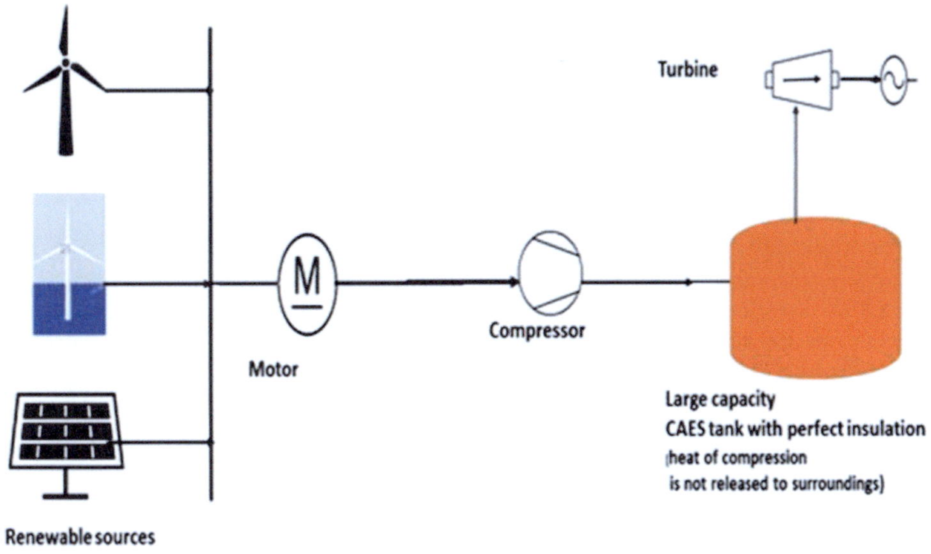

Fig. 2.3 Adiabatic compressed air energy storage system where the heat of compression is stored in the compressed air. Figure 1(b) from Venkataramani et al. (2018) Copyright (2018) The Authors. Reprinted with permission of Springer CC BY 4.0. https://creativecommons.org/licenses/by/4.0/

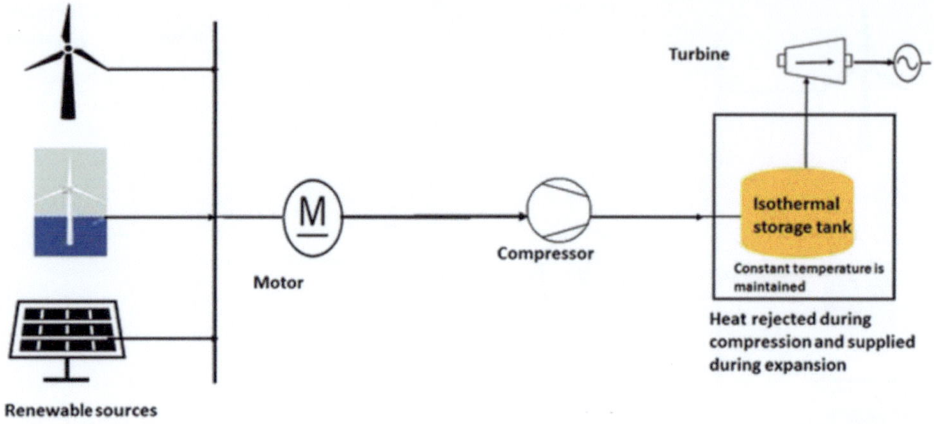

Fig. 2.4 Isothermal compressed air energy storage system where the compressed air tank is maintained at a constant temperature during energy storage and recovery. Figure 1(d) from Venkataramani et al. (2018) Copyright (2018) The Authors. Reprinted with permission of Springer CC BY 4.0. https://creativecommons.org/licenses/by/4.0/

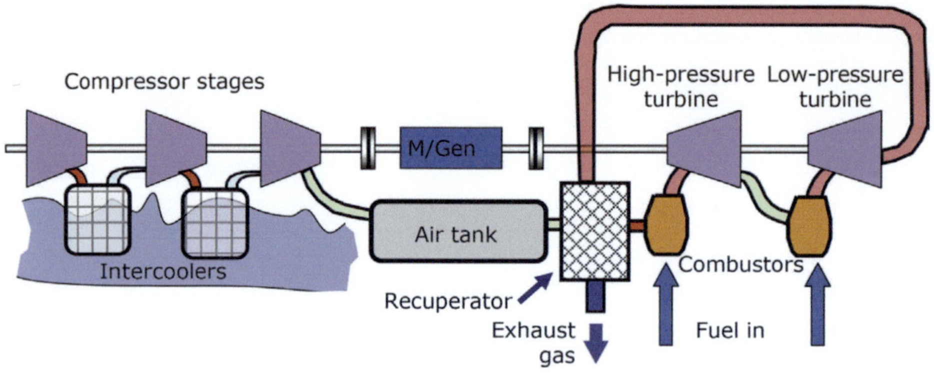

Fig. 2.5 A diabatic compressed air energy storage system using two-stage expansion and incorporating a recuperator. Figure 10 Reprinted from Garvey and Pimm (2016) Copyright (2016) with permission from Elsevier

such as natural gas, hydrogen and oil. Other possible sites include abandoned limestone mines.

Underwater compressed air energy storage requires the use of an appropriate containment vessel. Compressed air may be stored at a pressure comparable to the hydrostatic pressure. At a depth, d, in water the hydrostatic pressure is given by

$$P = \rho g d, \quad (2.11)$$

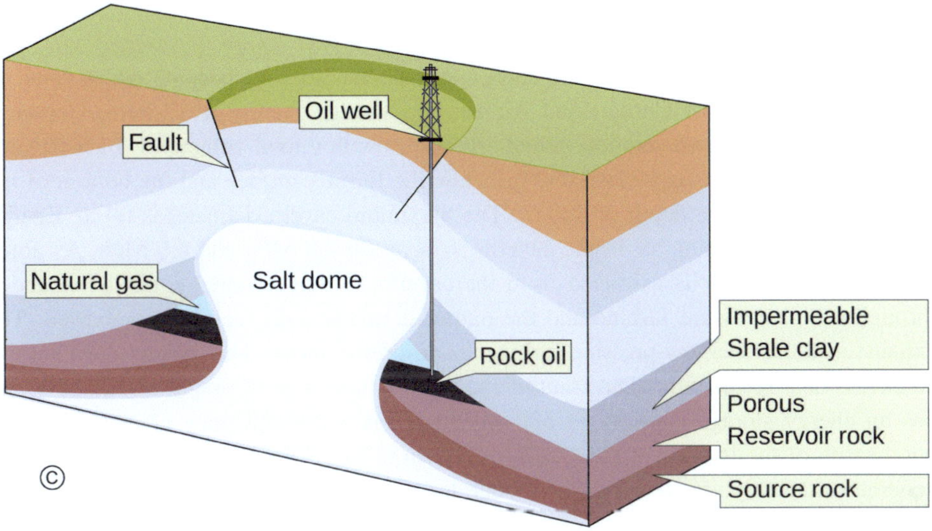

Fig. 2.6 Diagram of a salt dome. MagentaGreen (2014) CC BY-SA 3.0. https://creativecommons.org/licenses/by-sa/3.0/deed.en

where ρ is the water density and g is the gravitational acceleration. It is clear that the storage capacity will increase with water depth. Oceans provide the most obvious suitable locations, but deep-water lakes are also possibilities.

Vessels for underwater compressed air energy storage facilities fall into two general categories, flexible structures and rigid structures. Flexible vessels can be manufactured from air- and water-tight fabric. Energy is stored in the vessel by filling it with air at a pressure that is counteracted by the hydrostatic pressure of the surrounding water. Energy is recovered from the vessel by displacing the air with water. In the charged state the vessel will be subject to a buoyancy force which depends on the difference in density between the water and the compressed air, $(\rho_{water} - \rho_{air})$, as

$$F = (\rho_{water} - \rho_{air})gV, \tag{2.12}$$

where V is the vessel volume. The vessel must, therefore, be anchored appropriately to the sea (or lake) bottom. Rigid vessels constructed of (e.g.) concrete can incorporate sufficient ballast to counteract buoyant forces.

2.4 Use of Compressed Air Grid Storage

The first grid-scale compressed air energy storage facilities to become operational, in 1978, was the Huntorf Compressed Air Energy Storage facility in Elsfleth, Germany. This facility utilizes two solution-mined salt domes with a total volume of 3.1×10^5 m^3. The top of the caverns is located 650 m below the Earth's surface and the bottom of the caverns is located at a depth of 800 m. The maximum cavern diameter is 60 m. Energy is stored by pressurizing air in the caverns to between 4.8 MPa and 6.6 MPa. As noted above compressed air is extracted from the caverns to recover energy. Air is expanded through a high-pressure turbine and the output is mixed with fuel and combusted. The exhaust then drives a low-pressure turbine, as described above. This system does not use an excess heat recuperator. The Huntorf facility produces a peak output of 321 MW and has an energy storage capacity of 580 MWh. It has a net efficiency of about 42%. A photograph of the interior of the facility is shown in Fig. 2.7. Details of the compressor/generator assembly are shown in the model depicted in Fig. 2.8.

Fig. 2.7 Photograph of the interior of the Huntorf compressed air energy storage facility. From the left side of the image the units are high pressure compressor, transmission, low pressure compressor, motor/generator (see Fig. 2.8). Govgel (2006a) Public domain

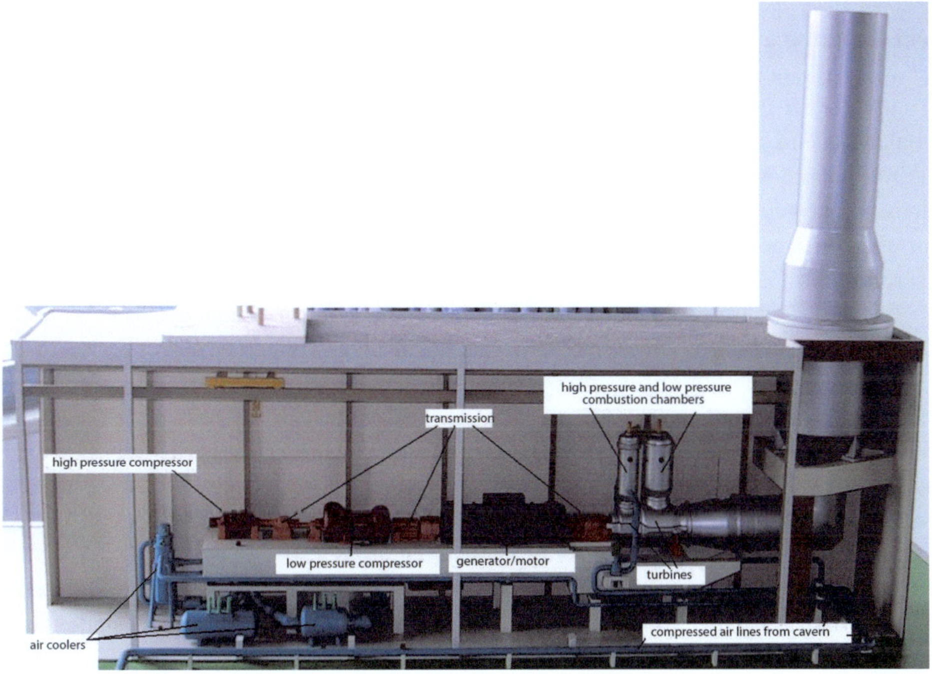

Fig. 2.8 Model of the interior of the powerhouse at the Huntorf compressed air energy storage facility in Huntorf near Elsfleth in Lower Saxony, Germany. Govgel (2006b) Public domain [labels translated to English]

In 1991 the world's second compressed air energy storage facility became operational in McIntosh, Alabama. This plant has a single cavern that was solution-mined from a salt dome. The top of the cavern is at a depth of 459 m and the bottom of the cavern is at a depth of 689 m. The maximum cavern diameter is 73 m and its total volume is 5.6×10^5 m^3. The facility has a peak output of 110 MW and a total energy storage capacity of 2,860 MWh. The McIntosh facility incorporates a heat recuperator, as illustrated in Fig. 2.5, and this approach reduces natural gas consumption by up to 25%, compared to the Huntorf facility, and increases the overall efficiency from around 42% to around 54%.

The design of the operational compressed air energy storage plants described above uses diabatic compression and, as a result, requires heating of the air by natural gas combustion during expansion. This approach uses fossil fuels and is, therefore, not a totally carbon-free energy storage approach. However, a detailed analysis suggests that these facilities use only 30% to 40% as much natural gas compared with a conventional natural gas turbine of the same energy output. More recent work on compressed air energy storage has sought to avoid the need for natural gas to heat the air during decompression.

Two of the approaches that have been considered are (1) to expand the gas isothermally and (2) to use a non-fossil fuel approach to provide the necessary heat.

A 2 MW near-isothermal compressed air energy storage facility operated in Gaines County, Texas from 2012 to 2016 (General Compression 2013; Evans 2022). The air was compressed in an underground cavern and the storage facility was coupled to a wind turbine in order to allow energy from the turbine to be stored during periods of low demand and release stored energy during periods of high demand. Since the compressed air cycle was near isothermal, no additional heat was needed during decompression and, therefore, no fossil fuels were used.

Recently two larger facilities have become operational in China, a 100 MW storage facility in Zhangjiakou and a 300 MW facility in Shangdong. The Zhangjiakou facility (Blain 2022; Robb 2023) became operational in 2022 and uses artificial storage vessels rather than a natural cavern for air storage. The system stores heat that is generated during air compression using supercritical thermal storage and recycles this heat using supercritical heat exchange to reheat the air as it is decompressed without the need for fossil fuel combustion (He et al. 2018). The storage facility has a total energy capacity of 400 MWh and an overall efficiency of 70.4%.

The 300 MW facility in Shangdong became operational in 2024 (Heat Exchanger World Publisher 2024; Murray 2024). Like the Zhangjiakou facility, the Shangdong storage facility uses supercritical thermal storage and supercritical heat exchange to achieve an overall efficiency of about 72%. However, unlike the Zhangjiakou facility, the Shangdong facility utilizes a natural salt cavern with depths up to about 1000 m. The overall volume of the cavern is more than 500,000 m^3 and provides a total energy storage capacity of 1.3 GWh.

Other similar energy storage projects using advanced compressed air energy storage techniques are being planned in the United States, Australia and Canada.

2.5 Other Applications of Compressed Air Energy Storage

While the use of compressed air energy storage for grid connected electricity is likely to be the most significant contribution of this technology to our energy systems, there have been other applications of compressed air in the past and these may also make contributions in the future. Specifically, compressed air has been considered as a storage mechanism for vehicle propulsion.

Compressed air has been in use to power locomotives for mining operations since the nineteenth century. For safety reasons, locomotives that utilize the combustion of a fuel for energy production cannot be used inside operating mines and compressed air has been one viable option. A notable example was the use of an air powered locomotive during the construction of the Gotthard Rail Tunnel in the 1870s and 1880s. This is a 15 km long tunnel that forms part of the Gotthardbahn, the railway that connects northern Switzerland to the southern Swiss Canton of Ticino. A more recent example of a compressed air mine locomotive is the Homestake Air Locomotive Number 1A used in the Homestake Gold Mine in Lead, South Dakota. This locomotive, illustrated in Fig. 2.9, was used from 1928 to 1961 to transport ore out of the mine shafts. The locomotive weighed over 12,000 kg and used compressed air at a pressure of 6.9 MPa in a 3.8 m^3 tank. The tank was filled from a large stationary air compressor. In addition to being a productive gold mine, the Homestake Mine was also the site of the Solar Neutrino Experiment conducted by Raymond Davis, Jr. (1914–2006) and John N. Bahcall (1934–2005) from 1970 to 1994. The mine ceased mining operations in 2001 and is currently the site of the Sanford Underground Laboratory used to detect the presence of Weakly Interacting Massive Particles (WIMPS).

The use of compressed air energy storage for road vehicles has also been considered. Tata Motors (Mumbai, India) announced that it planned to build a compressed air powered road vehicle (see Fig. 2.10). It was expected that the vehicle would include a source of combustion fuel that would heat the air during decompression on long trips. In 2017 they predicted that the vehicle would be available for sale by 2020 (Barooah 2017). At present it is not advertised on Tata's website and its future availability is not known.

Fig. 2.9 Homestake air locomotive number 1A used in the Homestake gold mine in lead, south Dakota from 1928 to 1961. Brossard (2017) CC BY-SA 2.0. https://creativecommons.org/licenses/by-sa/2.0/

Fig. 2.10 Tata/MDI CAT compressed air vehicle. Gupta (2007) CC BY-SA 4.0. https://creativecommons.org/licenses/by-sa/4.0/deed.en

In 2013 Peugeot Citroën announced that they would market a compressed air vehicle as early as 2016 (Carter 2013). It was intended to be based on the Citroën C3 and Peugeot 208. The vehicle would be a hybrid, using compressed air in additional to a conventional gasoline engine. The technology would follow traditional gasoline/battery hybrids with compressed air replacing the batteries. It was expected to use compressed air for speeds under 69 km/h and have a gasoline-free range of about 80 km. At present the vehicle is not yet available.

MDI (from Luxemburg) has advertised a compressed air urban vehicle (Grolms 2020), see Fig. 2.11. Their AirPod 2.0 is expected to have a maximum speed of 80 km/h and a range of 120 km. A version that also uses the combustion of a hydrocarbon-based fuel (e.g., bioethanol) has an extended range of 360 km. No vehicle has yet become available.

Following along the lines of the Chinese energy storage facilities, as described above, which utilize advanced methods to deal with heat requirements during diabatic decompression of the compressed air, recent research into compressed air vehicles has utilized the thermal properties of materials. Evrin and Dincer (2020), for example, at Ontario Tech University (Oshawa, Ontario, Canada) have developed a compressed air vehicle that uses

Fig. 2.11 MDI AirPod at the 2009 Geneva Motor Show. El monty (2009) CC BY-SA 3.0. https://creativecommons.org/licenses/by-sa/3.0/deed.en

phase change materials (PCM, see Chap. 6) to deal with heat recovery. Various phase change materials have been tested including polyethylene glycol, paraffin and an alkane mix. Paraffin has been found to perform the best. Their test on a prototype vehicle have shown a driving range of 140 km and an overall efficiency of about 60%, which is less than that of a comparable battery electric vehicle.

While compressed air is a possible alternative to other energy storage methods for vehicle applications, there is a major consideration to making this a viable option. Chemical fuels allow for the production of constant power. Up to the point where the fuel is exhausted. Batteries can also maintain reasonably constant power output throughout much of their discharge cycle. Compressed air, on the other hand, provides diminishing power as the air pressure decreases.

2.6 Combined Pumped Hydroelectric-Compressed Air Energy Storage

Another possible energy storage approach that deals with geographical constraints on traditional pumped hydroelectric storage and the need for natural gas combustion for traditional compressed air energy storage combines some aspects of both of these technologies. An illustration of a simple combined pumped hydroelectric-compressed air energy storage system is shown in Fig. 2.12. To store energy, water is pumped from the open tank on the left in the figure into the sealed air-pressurized chamber on the right. Energy is recovered by allowing water to run back through the pump/turbine into the open water tank. As the water is driven by air pressure not by differences in gravitational potential, there are no constraints on geography for the construction of such a system.

The detailed operation of a combined pumped hydroelectric-compressed air energy storage system is illustrated in Fig. 2.13. The high-pressure vessel (5) is pre-pressurized with air by compressor (4) through valve (8). The compressor is not used subsequently in the storage or recovery of energy. To store energy, water from the tank (3) is pumped into the high-pressure vessel through valve (7). To recover energy, water from the high-pressure vessel is forced by air pressure through valve (6) and the hydroelectric turbine (1) and back into the water tank. During energy storage and recovery, valves (9) and (10), respectively, are opened to spray water into the high-pressure vessel. This water exchanges heat with the air in the high-pressure vessel in order to reduce the effects of heating and cooling, during compression and expansion, respectively. This eliminates the need for natural gas combustion, as is typically used in commercial compressed air energy storage systems.

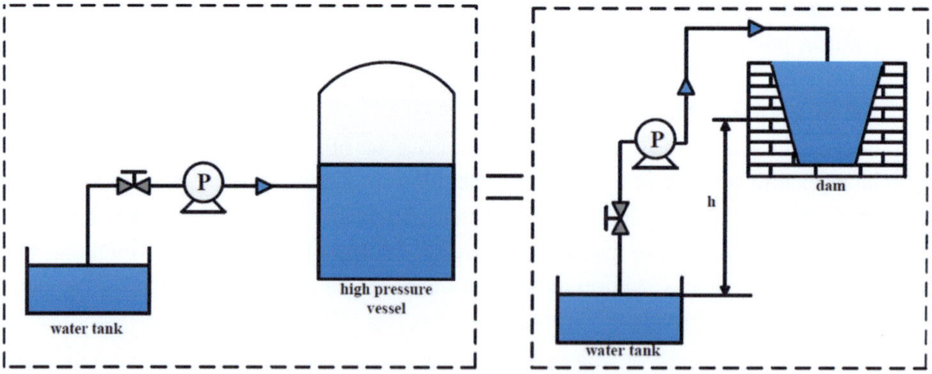

Fig. 2.12 Schematic of a simple combined pumped hydroelectric-compressed air energy storage system (*left*) and equivalent system based on gravitational potential (*right*). *P* pump/turbine. Figure 1 from Yao et al. (2015) Copyright (2015) by the Authors CC BY 4.0. https://creativecommons.org/licenses/by/4.0/

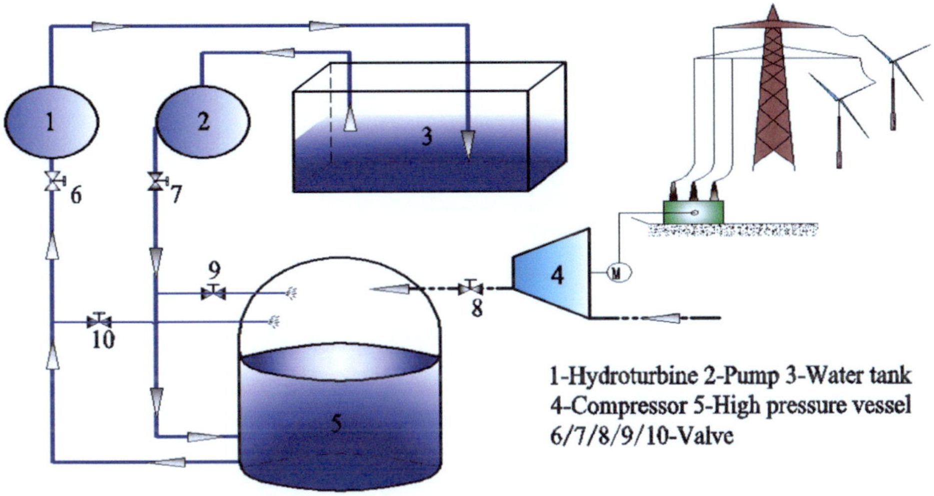

Fig. 2.13 Detailed diagram of the operation of a combined pumped hydroelectric-compressed air energy storage system. Figure 2 from Wang et al. (2013) Copyright (2013) by the Authors CC BY 3.0. https://creativecommons.org/licenses/by/3.0/

A major drawback of the energy storage system as described above is the fact that during energy storage and during energy recovery, the gas pressure in the high-pressure vessel does not remain constant. The consequence of this feature is that the energy storage rate (i.e. the power consumed) and energy recovery rate (i.e. the power generated) is not constant. This adverse characteristic of the system described above can be overcome by the design of the constant pressure system as shown in Fig. 2.14. In this system both the storage vessel and the high-pressure vessel are pre-pressurized by the compressor (C1) through valves (1) and (2), respectively. Compressor C1 is not subsequently used during energy storage or recovery. Energy is stored by pumping water from the water tank to the storage vessel through valves (3) and (4). Water is pumped into this vessel, compressor (C2) transfers air from the storage vessel to the high-pressure vessel through valve (5) so as to maintain the air pressure above the water in the storage vessel at a pre-set constant value. Energy is recovered by pressurized water from the storage vessel flowing through the turbine [through valves (6) and (7)]. As the water level in the storage vessel drops, compressed air is supplied from the high-pressure vessel through valve (8) in order to maintain the pressure in the storage vessel at a constant value.

A test system similar to the constant pressure design described above has been constructed by Camargos et al. (2018) and is shown in the diagram in Fig. 2.15. Pump (P-01) transfers water from the open water tank (TK-03) to the air/water tank (TK-02) for energy storage. A high-pressure Pelton turbine (T-01 in the figures) is used for recovery of energy from water flowing from the air/water tank (TK-02) to the open water tank (TK-03). The Pelton turbine is connected to a DC generator (GEN-01) to produce electricity.

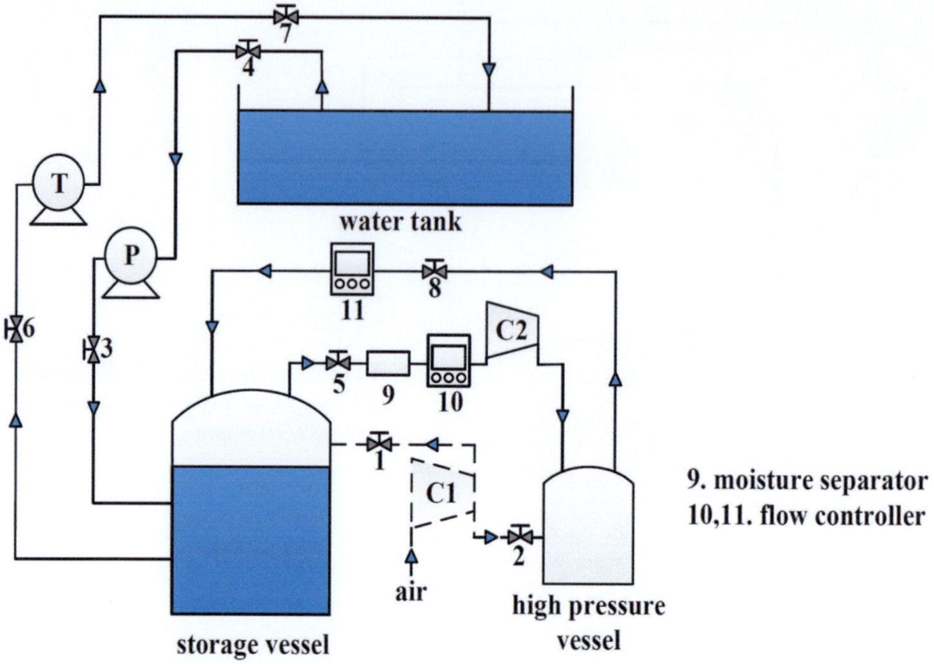

Fig. 2.14 Diagram of a constant pressure combined pumped hydroelectric-compressed air energy storage system; *P* pump, *T* turbine. Figure 2 from Yao et al. (2015) Copyright (2015) by the Authors CC BY 4.0. https://creativecommons.org/licenses/by/4.0/

Efficiencies comparable to those of commercial compressed air energy storage facilities have been obtained. The general concept utilized in pumped hydroelectric-compressed air energy storage systems should be readily scalable for systems with different capacities and applications.

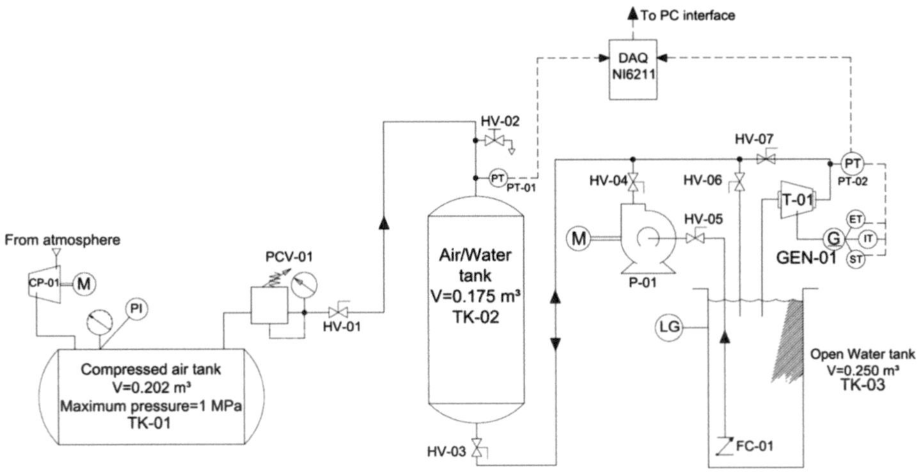

CP-01	Compressor	PI	Manometer
TK-01	Air tank	GEN-01	DC Generator
TK-02	Air/Water tank	P-01	Water pump
PCV-01	Pressure regulator	ST	Tachometer
PT1 and PT2	Pressure transducers	NI 6211	DAQ
T-01	Pelton turbine	FC-01	Foot valve
HV-01 to 07	Ball valves	LG	Level gauge
IT	Current sensor	TK-03	Open water tank

Fig. 2.15 Diagram of experimental combined pumped hydroelectric-compressed air energy storage system. Figure 2 Reprinted from Camargos et al. (2018) Copyright (2018) With permission from Elsevier

References

Barooah SB (2017) Tata motors' air-powered car project still on, to be launch ready in 3 years. Autocar Professional (14 Feb 2017). https://www.autocarpro.in/news-national/tata-motors-air-powered-car-project-launch-23659

Blain L (2022) China turns on the world's largest compressed air energy storage plant. New Atlas (04 Oct 2022) https://newatlas.com/energy/china-100mw-compressed-air/

Brossard B (2017) Air Loco. https://www.flickr.com/photos/string_bass_dave/37287547392/

Camargos TPL, Pottie DLF, Ferreira RAM, Maia TAC, Porto MP (2018) Experimental study of a PH-CAES system: proof of concept. Energy 165:630–638. https://doi.org/10.1016/j.energy.2018.09.109

Carter M (2013) Peugeot announces plans to release a hybrid car that runs on compressed air by 2016. INHABITAT (25 Jan 2013). https://inhabitat.com/peugeot-announces-plans-to-release-a-hybrid-car-that-runs-on-compressed-air-by-2016/

El monty (2009) MDI AirPod at the 2009 Geneva motor show. https://commons.wikimedia.org/wiki/File:MDI_Air_Pod_(1).JPG

Evans DJ (2022) The geology, historical background and developments in CAES, chap. 18. In Hauer A (ed) Advances in energy storage—latest developments from R&D to the market. Wiley, Hoboken, NJ, pp 323−389

Evrin RA, Dincer I (2020) Experimental investigation of a compressed air vehicle prototype with phase change materials for heat recovery. Energy Storage 2:e159. https://doi.org/10.1002/est2.159

Garvey SD, Pimm A (2016) Compressed air energy storage, chap 5. In Letcher TM (ed) Storing energy—with special reference to renewable energy sources. Elsevier, Amsterdam, pp 87−111. https://doi.org/10.1016/B978-0-12-803440-8.00005-1

General Compression (2013) Texas dispatachable wind. https://web.archive.org/web/20130523153848/http://www.generalcompression.com/index.php/tdw1

Govgel (2006a) Kraftwerk von innen mit Verdichtern im Vordergrund. https://commons.wikimedia.org/wiki/File:Kraftwerk_Huntorf_innen.jpg

Govgel (2006b) Modell vom Kraftwerk Huntorf. https://commons.wikimedia.org/wiki/File:Kraftwerk_Huntorf_Modell.jpg

Grolms M (2020) Compressed air cars for urban transportation. Adv Sci News (7 Sept 2020) https://www.advancedsciencenews.com/compressed-air-cars-for-urban-transportation/

Gupta D (2007) Tata/MDI CAT compressed air car. https://commons.wikimedia.org/wiki/File:Catvertroquette.jpg

He Q, Liu H, Hao Y, Liu Y, Liu W (2018) Thermodynamic analysis of a novel supercritical compressed carbon dioxide energy storage system through advanced exergy analysis. Renew Energy 127:835–849. https://doi.org/10.1016/j.renene.2018.05.005

Heat Exchanger World Publisher (2024) World's largest compressed air energy storage project comes online in China. Heat Exchanger World (27 May 2024). https://heat-exchanger-world.com/worlds-largest-compressed-air-energy-storage-project-comes-online-in-china/

MagentaGreen (2014) Scheme of an oil trap on salt dome flanks. https://commons.wikimedia.org/wiki/File:Salt_dome_trap.svg

Murray C (2024) World's largest compressed air energy storage project connects to the grid in China. Energy Storage News (10 Apr 2024). https://www.energy-storage.news/worlds-largest-compressed-air-energy-storage-project-connects-to-the-grid-in-china/

Rabi AM, Radulovic J, Buick JM (2023) Comprehensive review of compressed air energy storage (CAES) technologies. Thermo 2023:104−126. https://doi.org/10.3390/thermo3010008

Robb A (2023) Can compressed air energy storage solve the long-duration dilemma? Renew Energy World (12 Apr 2023). https://www.renewableenergyworld.com/storage/can-compressed-air-energy-storage-solve-the-long-duration-dilemma/#gref

Venkataramani G, Ramalingam V, Viswanathan K (2018) Harnessing free energy from nature for efficient operation of compressed air energy storage system and unlocking the potential of renewable power generation. Sci Rep 8:9981. https://doi.org/10.1038/s41598-018-28025-5

Wang H, Wang L, Wang X, Yao E (2013) A novel pumped hydro combined with compressed air energy storage system. Energies 6:1554–1567. https://doi.org/10.3390/en6031554

Yao E, Wang H, Liu L, Xi G (2015) A novel constant-pressure pumped hydro combined with compressed air energy storage system. Energies 8:154–171. https://doi.org/10.3390/En8010154

Energy Storage Using Solid Masses 3

3.1 Introduction

The mechanical energy storage methods described in Chaps. 1 and 2 involved the use of liquids and gases, respectively. These methods utilized the gravitational potential of water, in the first case, and the energy associated with compressed air in the second. It is also possible to store energy using a solid mass, either as gravitational potential energy or as rotational kinetic energy. The present chapter reviews the method for utilizing these approaches.

3.2 Gravitational Potential of Solid Masses

Following along the lines of pumped hydroelectric storage, it has been suggested that solid masses may be raised and lowered to store energy by making use of the differences in gravitational potential. The energy storage capacity is described by

$$E = \eta m g \Delta h, \tag{3.1}$$

where η is the net system efficiency, m is the mass, g is the gravitational acceleration and Δh is the difference in height. It is clear that the energy storage capacity is maximized by maximizing the mass and the change in height; although, from a practical standpoint, there may be limits to either or both of these quantities. In the present section we look at four possible approaches to this method of storing energy.

R. A. Dunlap, *Renewable Energy Storage*, Synthesis Lectures on Renewable Energy Technologies, https://doi.org/10.1007/978-3-031-88942-4_3

3.2.1 Advanced Rail Energy Storage

One viable approach to raising and lowering solid masses to store energy is a rail-based system for moving masses up and down a long incline. Advanced Rail Energy Storage (ARES) LLC has tested a prototype system of this type in Tehachapi, California (near a major wind farm) as shown in Fig. 3.1 (Cava et al. 2016). The rail vehicle containing an electric motor/generator is loaded with large masses and travels along a track. Additional non-motorized cars may be attached to the motorized vehicle to increase mass. Typical masses in the order of 200 t would be used. Electrical connection for the motor/generator is provided by a third rail, as is utilized for electric trains.

As with pumped hydroelectric storage facilities, appropriate geography is necessary for a rail energy storage system. A grade of less than about 10% is needed for the rail drive system to function, so a long gradual incline is necessary. An incline of 8% over a distance of 15 km would give a total difference in height of 1200 m, corresponding to a stored energy of about 0.5 MWh per (200 t) vehicle. Note that the necessary geography is quite different than that required for pumped hydroelectric storage where typically a 100% grade is used for penstocks.

Fig. 3.1 Prototype rail energy storage vehicle in Tehachapi, California. Note the wind farm along the horizon. Figure 4.7 Reprinted from Cava et al. (2016) Copyright (2016) with permission from Elsevier

Such a concept is scalable, within a fairly wide range of sizes. Capacity might range from about 25 MW with 6 MWh storage (for an output duration of about 15 min) to 2000 MW with 240 GWh storage (for an output duration of 120 h).

Although such an approach to energy storage is still at its early stages of technical development, there are a number of possible advantages over the commonly used pumped hydroelectric storage system. These include.

- about half the initial infrastructure cost
- scalable over a larger range of sizes
- potentially greater availability of locations with appropriate geography
- minimal post-lifetime environmental impact
- excellent net efficiency of around 90%.

3.2.2 Gravity Power Module

A novel approach to the use of pumped hydroelectric storage combined with gravitational energy is being developed by Gravity Power LLC (Gravity Power LLC ND). Their Gravity Power Module (GPM) consists of a vertical cylindrical shaft cut into the Earth containing a piston which can move vertically in the shaft (Galant et al. 2013). Typically, the piston is a composite made of reclaimed rock from the excavation along with concrete. The shaft above and below the piston is filled with water. Figure 3.2 shows the principle of operation. Energy is stored by pumping water from above the piston to the space below the piston, as shown in Fig. 3.2a, thereby raising the piston. Energy is generated by allowing the piston to fall, thereby forcing water from below the piston to pass through the generator to the space above the piston as shown in Fig. 3.2b. The total energy storage capacity follows from Eq. (3.1) and can be written as

$$E = \eta\left(\rho_{piston} - \rho_{water}\right) g V_{piston} \Delta h \tag{3.2}$$

where η is the net efficiency, $(\rho_{piston} - \rho_{water})$ is the difference in density between the piston and the water, g is the gravitational acceleration, V_{piston} is the volume of the piston and Δh is the change in vertical position of the center of mass of the piston. The dimensions of this device can be scaled, but a typical module might consist of a 500 m deep shaft, 30 m in diameter containing a 250 m long piston. The energy storage capacity of such a device is obtained using typical values of the parameters in Eq. (3.2); $\eta = 0.8$ and $\rho_{piston} = 2500\ \text{kg/m}^3$. So, for $V_{piston} = \pi(15\ \text{m})^2 \times 250\ \text{m} = 1.77 \times 10^5\ \text{m}^3$ and $\Delta h = 250\ \text{m}$, then

$$\begin{aligned} E &= (0.8) \times (2500 - 1000)\text{kg/m}^3 \times 9.8\ \text{m/s}^2 \\ &\times 1.77 \times 10^5\ \text{m}^3 \times 250\ \text{m} - 5.21 \times 10^{11}\ \text{J} \end{aligned} \tag{3.3}$$

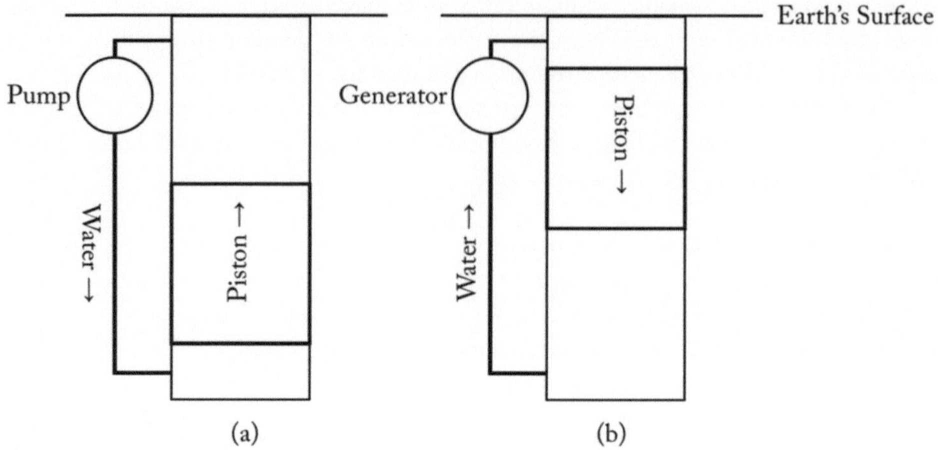

Fig. 3.2 Concept of the Gravity Power Module; **a** energy storage and **b** energy generation

or 145 MWh. This can represent a maximum power of 40 MW for a duration of about 3.6 h.

While a full-scale device of this type has not been constructed, a MW-size demonstration project under development.

3.2.3 GraviStore

As noted in Chaps. 1 and 2, abandoned mines are attractive locations for possible pumped hydroelectric energy storage and compressed air energy storage. First Quantum Mineral's Pyhäsalmi Mine, in the town of Pyhäjärvi, Finland is the deepest zinc, copper and pyrite mine in Europe. It began operations in 1962 and has a maximum depth of 1450 m. Like many other mines, it has found an applications as suitable environment for particle physics experiments as shown in Fig. 3.3 (Enqvist 2006).

Since underground mining operations at the Pyhäsalmi Mine ceased in the early 2020s, the mine has been considered as a suitable location for renewable energy storage technologies. It was proposed that the very large head that could be obtained from the deep mine shafts would be ideal for pumped hydroelectric storage (Callio 2024). However, plans are underway to utilize one of the mine shafts for gravity storage (Rudge 2024; Staff Writer 2024; Tisheva 2024).

Energy company Gravitricity from Edinburgh, Scotland (Gravitricity 2024) is developing a gravitational based energy storage system named Gravistor in the Pyhäsalmi Mine. This system, sometimes called a gravity battery, uses a weight that is raised or lower to store and release energy, respectively, as shown in Fig. 3.4.

Fig. 3.3 EMMA (Experiment with MultiMuon Array) cosmic ray experiment inside the Pyhäsalmi Mine, in the town of Pyhäjärvi, Finland. Sarffman (2013) CC BY-SA 3.0. https://creativecommons.org/licenses/by-sa/3.0/deed.en

The GraviStore system has been demonstrated using an above-ground system consisting of a 12 m high test rig and two 25 t weights. The system generates 250 kW and reaches maximum power output within one second of releasing the weight. A full-scale prototype which would produce up to 2 MW is planned for construction in a 530 m auxiliary shaft in the Pyhäsalmi Mine.

Gravitricity is also working on similar energy storage systems in other European mines, including the Velenje coal mine near Velenje, Slovenia, the Darkov mine near Ostrava in the Moravian-Silesian region of the Czech Republic and the Grube Teutschenthal mine near Halle, Germany.

3.2.4 Energy Vault

Energy Vault is an energy storage company located in Lugano, Switzerland (Energy Vault 2024). One of Energy Vault's development projects (shown in Fig. 3.5) utilizes the gravitational potential of concrete blocks raised by the aid of a crane. The prototype shown in the figure was constructed in 2019 and uses a 70 m tower with three double cranes attached. The concrete blocks are each about 5 m high and weigh 35 t. Blocks are stacked,

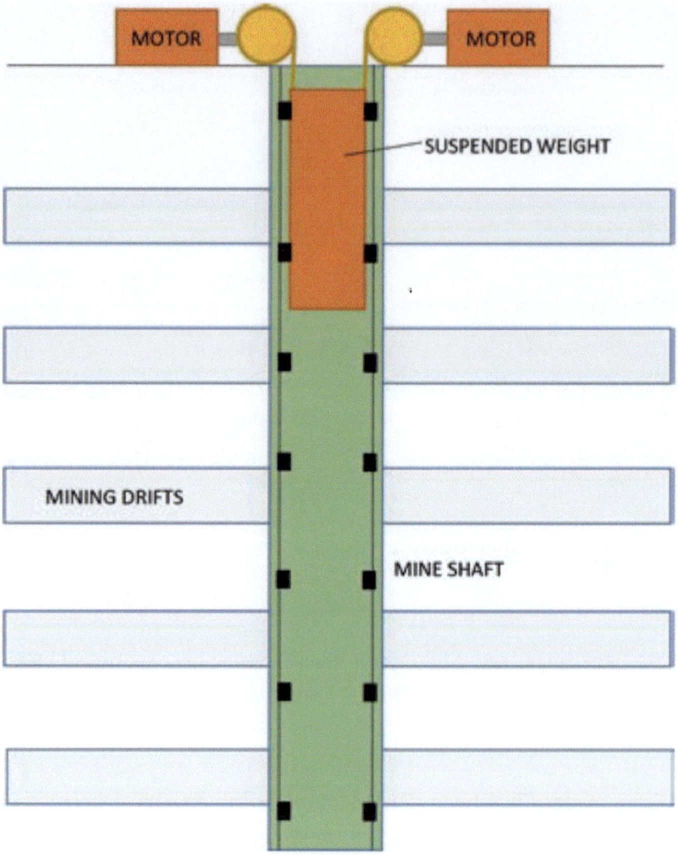

Fig. 3.4 Diagram of a gravitational energy storage system (gravity battery) in a mine shaft. Figure 3 from Menendez et al. (2020) CC BY 4.0. https://creativecommons.org/licenses/by/4.0/

as shown in the figure, with a maximum of eight layers of blocks in this prototype device. The figure shows blocks stacked in two piles, one on either side of the tower up to four layers high. According to Eq. (3.1) each block stores a maximum energy of

$$\left(35 \times 10^3\,\mathrm{kg}\right) \times \left(9.8\,\mathrm{m/s^2}\right) \times 5\,\mathrm{m} = 1.7\,\mathrm{MJ} \tag{3.4}$$

or about 0.47 kWh. Energy is recovered by lowering the block and driving a generator using the cable connected to the block. The round-trip efficiency is expected to be about 90%. Additional blocks, to those shown in the figure, can be added in two concentric circular piles constructed around the tower.

Fig. 3.5 Energy Vault Testing Tower (EV1) in Castione-Arbedo. Switzerland. Keimzelle (2022) CC BY-SA 4.0. https://creativecommons.org/licenses/by-sa/4.0/deed.en

A full size Energy Vault storage device would utilize up to 7000 blocks of 35 t each, of which about 4300 could be moved vertically with an average displacement of 90 m. This configuration would provide a maximum energy storage capacity of about 38 MWh.

3.3 Flywheels

Flywheels are axially symmetric devices designed to store rotational energy and have been in use since antiquity (e.g., for potter's wheels). Applications of flywheels include:

- Smoothing the rotational output of devices that convert reciprocating motion to rotary motion. This was probably the first industrial use of the flywheel for steam engines developed in the late eighteenth century, see Fig. 3.6. Flywheels on large nineteenth century stationary steam engines could be up to 9 m in diameter.

Fig. 3.6 Hick Hargreaves No. 303 steam engine constructed at Bolton, Lancashire, U.K. in 1873. This steam engine has a single cylinder of 33 cm diameter with a 91 cm stroke and produces 37 kW. The flywheel has a diameter of about 3 m and rotates at 66 rpm. Dace (2010) CC BY-SA 2.0. https://creativecommons.org/licenses/by-sa/2.0/deed.en

- Controlling or measuring the orientation of a device by use of a gyroscope. These devices were developed around the beginning of the nineteenth century.

More recently, flywheels have been considered as a means of storing energy. Specifically, electrical energy can be used to rotate the flywheel by means of an electric motor and the energy can be recovered by slowing the flywheel as it turns a generator. A simple system for using a flywheel to store electrical energy is shown in Fig. 3.7.

3.3.1 The Physics of Flywheels

The rotational kinetic energy of an object can be calculated by beginning with the translational kinetic energy of a mass, m, moving with a velocity, v, as

$$E = \frac{1}{2}mv^2. \tag{3.5}$$

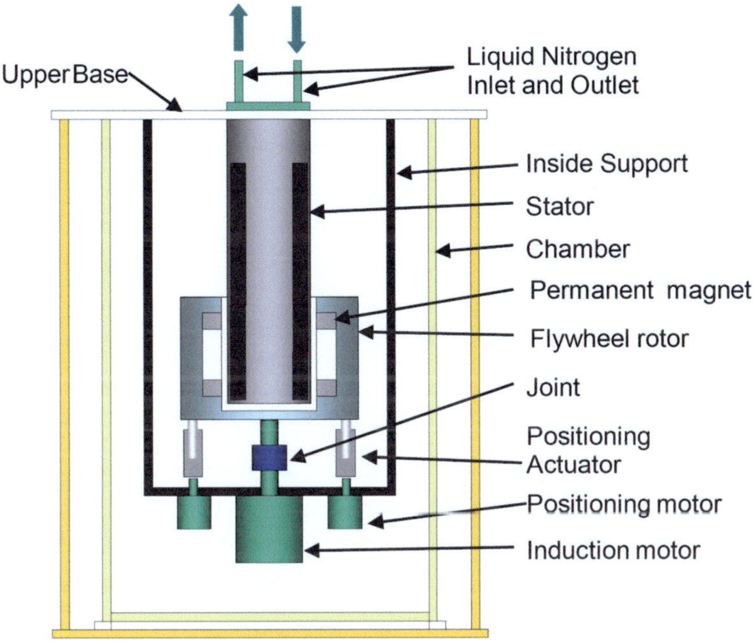

Fig. 3.7 Typical design of a flywheel energy storage system. Figure 1 from Komori et al. (2022) Copyright (2022) by the Authors CC BY 4.0. https://creativecommons.org/licenses/by/4.0/

If we consider a differential mass, *dm*, as shown in Fig. 3.8, moving with a velocity *v*, the kinetic energy will be

$$dE = \frac{1}{2}v^2 dm. \tag{3.6}$$

If the perpendicular distance from the differential mass to the *z*-axis is *r*, then the angular velocity about the *z*-axis will be

$$\omega = \frac{v}{r} \tag{3.7}$$

and Eq. (3.6) may be written as

$$dE = \frac{1}{2}\omega^2 r^2 dm. \tag{3.8}$$

If we now consider the rotational kinetic energy of a distributed object about the *z*-axis, we can integrate its differential mass elements as

$$dE = \frac{1}{2}\omega^2 \iiint r^2 dm, \tag{3.9}$$

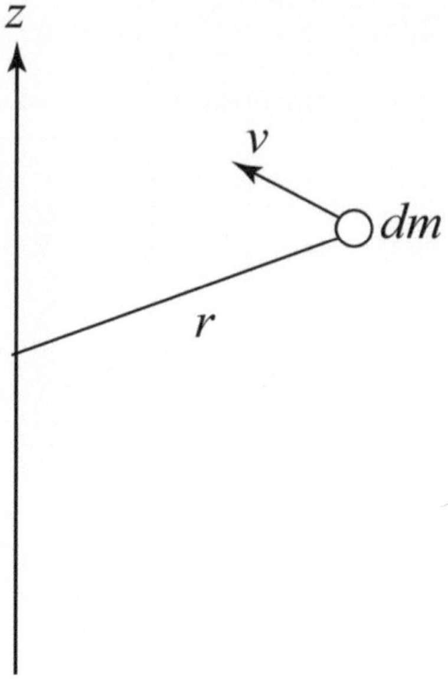

Fig. 3.8 Kinetic energy of a differential mass moving with a velocity v at a perpendicular distance r from the z-axis

where the integral is over all space and is a function of the geometry and mass distribution of the object. This integral is defined as the moment of inertia of the object, I, and Eq. (3.9) can be written in its usual form as

$$E = \frac{1}{2} I \omega^2. \tag{3.10}$$

The moment of inertia from Eq. (3.9) may be written as an integral of the density distribution of the object over its volume,

$$I = \iiint \rho(r) \mathrm{r}^2 dV, \tag{3.11}$$

where $\boldsymbol{r}$ is the vector defining the perpendicular distance between the volume element and the z-axis.

The integral in Eq. (3.11) can be evaluated for masses of different geometries. In the simplest case we can consider different geometries for objects of total mass, m, that have constant density and are axially symmetric about the z-axis (axis of rotation) with a total radius, R. In this case, the moment of inertia from Eq. (3.11) can be expressed as

$$I = kmR^2, \tag{3.12}$$

where the constant k is determined by the object's geometry. Equation (3.12) can be combined with Eq. (3.10) to give

Table 3.1 The constant k for the moment of inertia in Eq. (3.13) for some axially symmetric geometries

Geometry	k
Solid sphere	2/5
Disk	1/2
Cylinder	1/2
Spherical shell	2/3
Ring	1

$$E = \frac{1}{2} kmR^2\omega^2. \tag{3.13}$$

Values of k for some common geometries are given in Table 3.1. For a simple flywheel, it can be seen from Eq. (3.13) and the values of k in Table 3.1, that the maximum energy can be stored in a ring-shaped flywheel, as this geometry places as much of the mass as far from the axis of rotation as possible. Such a geometry is sometimes referred to as a rim-loaded flywheel as shown in Fig. 3.6. There are, however, limits to this approach for energy storage because of the internal stresses in flywheels of different geometries. From a practical standpoint, as discussed below, a thick-walled cylinder (or even a disk) is generally more appropriate than a thin rimmed cylinder. A thick-walled cylinder is defined by an inner radius of R_i and an outer radius of R_o. The moment of inertia of this geometry is given by

$$I = \frac{1}{2} m\left(R_i^2 + R_o^2\right) \tag{3.14}$$

One might expect that materials such as iron, as has traditionally been used for flywheels (e.g., Fig. 3.6), would be the most suitable for storing large amounts of energy. However, a careful analysis of the properties of different materials shows that this is not necessarily the case. The rim of the flywheel has the highest velocity and is, therefore, subject to the greatest stress. It can be shown that the rim stress is given by

$$\tau = \rho R^2\omega^2, \tag{3.15}$$

where ρ is the bulk density of the material. The maximum allowable stress is the breaking stress of the flywheel material (generally taken to be the tensile strength, $\tau = \sigma$). Combining Eq. (3.15) with Eq. (3.13) for the energy gives the maximum energy that can be stored as

$$E = \frac{1}{2} km\left(\frac{\sigma}{\rho}\right), \tag{3.16}$$

where E is in J when mass is in kg, σ is in N/m^2 and ρ is in kg/m^3. The maximum energy storage capacity of a flywheel is, therefore, related to its geometry (through k), its mass, m, and the intrinsic material property (σ/ρ). Since the maximum energy is linearly related to m and (σ/ρ), choosing the material with the highest possible value of (σ/ρ) should yield the flywheel with the highest possible energy storage capacity per unit mass. It is interesting to note that the units of (σ/ρ) are N m/kg = J/kg and that all other factors being equal, low-density materials are preferable to high density materials.

Another way of looking at Eq. (3.16) is to use the relation mass/density = volume to obtain

$$E = \frac{1}{2} k \sigma V. \tag{3.17}$$

Note that the units of tensile strength are N/m^2 = J/m^3 and that the material with the highest tensile strength will yield the flywheel with the highest energy capacity per unit volume.

3.3.2 Flywheel Design Criteria

While the discussion above provides some general guidance to the design of a flywheel, there are a number of specific factors that need to be considered in some detail in order to optimize its functionality and efficiency. These factors include the following:

- the flywheel material
- the flywheel geometry
- the flywheel enclosure
- the motor/generator
- the support of the flywheel (i.e. the bearings)

These are discussed briefly below.

3.3.2.1 Flywheel Materials

Table 3.2 gives the properties of some materials that may be considered for flywheel construction. It is clear that the use of cast iron as a flywheel material, as was traditional for steam engines in the nineteenth century, is not suitable for energy storage purposes. The table also shows that materials that we typically think of as "strong" are not necessarily the best choice because of their high density. Carbon fiber composites are seen to offer the best option for energy storage, both in terms of energy storage capacity per unit mass (highest σ/ρ) and per unit volume (highest σ).

Table 3.2 Properties of some potential flywheel materials

Material	Density (ρ) (kg/m^3)	Tensile strength (σ) (N/m^2)	σ/ρ (N m/kg)
Cast iron	7870	0.2×10^9	0.025×10^6
Magnesium alloy	1740	0.24×10^9	0.14×10^6
Steel	7870	1.72×10^9	0.22×10^6
Aluminum alloy	2700	0.59×10^9	0.22×10^6
Beryllium	1850	0.48×10^9	0.26×10^6
Titanium alloy	4500	1.22×10^9	0.27×10^6
Fiberglass	2000	1.60×10^9	0.80×10^6
Carbon fiber composite	1500	2.40×10^9	1.60×10^6

The energy storage capacity per unit mass is related to σ/ρ and the energy storage capacity per unit volume is related to σ. *Note* values are typical for each class of material, as tensile strength can vary considerable from one alloy to another

3.3.2.2 Flywheel Rotor Geometries

It is seen from the above discussion that a rim-loaded flywheel design provides the highest moment of inertia because it maximizes the distance of the mass from the axis of rotation. However, the rim of the flywheel is traveling at the greatest velocity and, therefore, has the greatest stress. Distributing some of the mass of a flywheel rotor closer to the axis of rotation (compared to a rim-loaded design), will decrease the overall moment of inertia, but it may increase the total energy storage capacity by reducing the stress at the rim and allowing greater rotational speeds. A commonly used approach to analyzing different flywheel rotor shapes is to rewrite Eqs. (3.16) and (3.17) in terms of the maximum energy per unit mass and per unit volume as

$$\frac{E}{m} = K\left(\frac{\sigma}{\rho}\right) \tag{3.18}$$

and

$$\frac{E}{V} = K\sigma, \tag{3.19}$$

where K is a shape factor that is determined by the rotor geometry. Figure 3.9 illustrates the values of K for some different geometries. The so-called *Disk of Laval* is the shape that has constant stress across the rotor and is also the shape that has the highest value of K. Different flywheels may utilize differently shaped rotors in order to optimize the design for a particular application. A common approach to rotor design is the use of a composite flywheel where the inner portion of the flywheel is made from a material such as steel or titanium and the rim of the flywheel, which experiences the greatest stress, is made of carbon fiber.

Shape	Cross Section	K
Laval disk		1.00
Laval disk real		0.70–0.90
Conical disk		0.70–0.85
Solid disk		0.606
Thin ring	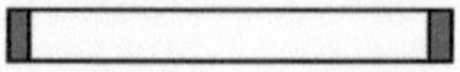	0.50
Thick rim	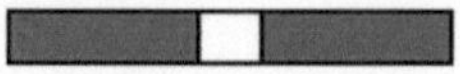	0.303

Fig. 3.9 Shape factors for some different flywheel rotor geometries. Table 2 from Skinner and Mertiny (2022) Copyright (2022) The Authors CC BY 4.0. https://creativecommons.org/licenses/by/4.0/ (k-value changed to K)

3.3.2.3 Flywheel Enclosures

In order to reduce energy loss due to friction between the flywheel rotor and air, it is customary to enclose the flywheel in a vacuum chamber, as shown in Fig. 3.7. This enclosure is typically made of a structural material such as steel. In some applications where energy losses are not a major concern, flywheel enclosures may be filled with low-pressure helium.

3.3.2.4 Motor/Generator

In most cases the motor/generator is mounted inside the vacuum enclosure and is coupled directly to the flywheel rotor, as shown in Fig. 3.7. In some designs the motor/generator may be outside of the vacuum enclosure and may be magnetically coupled to the flywheel shaft. There are a number of different types of motor/generators that can be used for flywheel energy storage, depending on the specific application and power levels involved. One major concern that needs to be addressed with regard to the choice of the motor/generator is that of heat. Motor/generators produce heat during the charging and discharging of the flywheel. In systems which are used (e.g.) for backup power, the flywheel is charged or discharged only rarely, and heating is generally not an issue. In systems which are cycled often, heating can be an important factor. Heat is conducted through the shaft to the flywheel but, because the system is enclosed in a vacuum, heat is dissipated very slowly. For metallic flywheels (e.g. steel or titanium) heating is generally not a problem, however, for composite flywheels (as discussed in the applications section) excess temperatures can be detrimental to the rotor. In such cases, motor/generators which generate as little heat as possible are desirable and cooling, as indicated in Fig. 3.7, may be included in the assembly. Permanent magnet motors are generally the best choice for flywheels designed for energy storage.

3.3.2.5 Bearings

Bearings support the flywheel rotor and allow it to rotate freely, with as little friction as possible. Generally, stand alone flywheel systems are configured with a vertical axis so that bearing loads are distributed symmetrically. Traditionally, mechanical bearings (i.e. balls in races) have been used for rotating machinery. Since flywheels that are designed to store significant amounts of energy are heavy and rotate at high speeds, bearings must be appropriately designed. In most cases, bearing life is the limiting factor for continuous flywheel operation. It should also be noted that the usual vacuum requirements for flywheel operation place additional demands on heat dissipation and lubrication.

Most advanced flywheel systems utilize magnetic bearings, sometimes referred to as active magnetic bearings (AMBs), where the flywheel rotor is levitated in a magnetic field. A common configuration uses permanent magnets to levitate the rotor assembly and small electromagnets to provide fine positioning and stability. Magnetic bearings typically introduce less energy loss than mechanical bearings. Other advantages of magnetic bearings result from the fact that the rotating flywheel assembly is not in physical contact with the stationary portions of the flywheel enclosure. This means that heat generation in the bearings is not an issue and no lubricant is needed. The lack of lubricant benefits the production of the vacuum environment. One disadvantage of magnetic bearings, however, is the need for so-called touch-down bearings, which support the flywheel in the event of a magnetic bearing failure. Magnetic bearings may also be used in conjunction with mechanical bearings in order to minimize bearing load and wear. Superconducting levitation systems based on the Meissner effect may also be used to support a flywheel rotor. Such systems are in the early stages of commercial development.

3.3.3 Applications of Flywheels for Energy Storage

3.3.3.1 Grid Storage and Stabilization

The design of a typical flywheel energy storage system for grid backup and stabilization is illustrated in Fig. 3.10. A number of such systems are now commercially available. A popular example is the Model A32 from Amber Kinetics (2024). This system has an energy storage capacity of 32 kWh and a maximum output of 8 kW (for up to 4 h). The round-trip efficiency is greater than 85%. While the rated capacity is much less than pumped hydroelectric storage or proposed systems involving the gravitational potential of solid masses, flywheel energy storage modules have found applications for grid stabilization and for covering short-term power interruptions. A large number of flywheel modules can be connected together to increase energy storage and power capacity. For example, a grid storage system was installed in Stephentown, New York by Beacon Power (Power Technology 2021), consisting of 200 flywheel modules for a total energy storage capacity of 5 MWh and a maximum power output of 20 MW. Typical units have a self-discharge rate of about 5% per day but can be kept fully charged from the grid until needed.

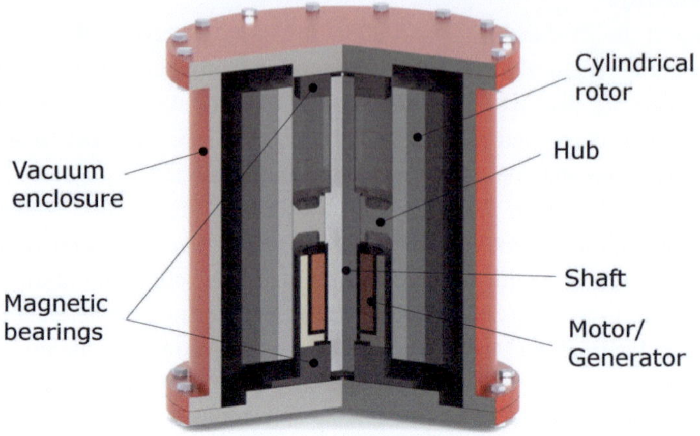

Fig. 3.10 Typical design of a flywheel energy storage system used for grid storage and stabilization. Pjrensburg (2012) CC BY-SA 3.0. https://creativecommons.org/licenses/by-sa/3.0/deed.en

3.3.3.2 Vehicle Propulsion

Flywheels may provide an alternative means of vehicle propulsion. Here we consider the suitability of flywheel energy storage capacity for such an application. We consider the maximum specific energy storage for a carbon fiber composite flywheel as given in Table 3.2 of 1.6 MJ/kg (although one may choose a smaller value for safety). If we assume a total flywheel mass of 100 kg, then the total energy storage will be 160 MJ. A midsize passenger vehicle typically requires an average of about 0.55 MJ of energy at the wheels per km of distance travelled leading to a range for the flywheel vehicle of (160 MJ/0.55 MJ/km) = 290 km, similar to that of many battery electric vehicles. In the transportation sector, flywheels have most commonly been used as a means of propulsion for city buses (see Figs. 3.11 and 3.12). In this case, weight, size and/or cost considerations may not be as important as for personal passenger vehicles.

It may also be possible to construct hybrid vehicles which use a flywheel in conjunction with another source of energy. Flywheels have the advantage of greater specific power than batteries and are a very suitable method of storing energy from regenerative braking. This approach has been successfully applied to racing cars, such as the Le Mans Prototype 1 (LMP1) Audi R18 e-tron quattro shown in Fig. 3.13.

3.3.3.3 Transit System Energy Recovery

Flywheels may be used to recover lost energy from electric trains used for public transit. A flywheel energy storage system may be installed at a transit station and energy produced during regenerative braking when the train stops at the station can be stored in the flywheel. This energy may then be used to provide power to the train for acceleration when it leaves the station. This approach recovers energy that would otherwise be lost

Fig. 3.11 Gyrobus G3 flywheel bus built in 1955 on exhibit in the Flemish Tramway, Trolleybus and Bus Museum, Antwerp, Belgium. Volkov (2008) CC BY 2.0. https://creativecommons.org/licenses/by/2.0/deed.en

and helps to reduce voltage sag on the transit power line, which can occur when additional load is placed on the line by accelerating trains. Demonstration systems have been installed at transit stations in Los Angeles, New York, London, Lyon and Paris and have provided positive evidence for the utility of this approach. (see e.g., London Midland 2008).

3.3.3.4 Spacecraft Energy

NASA has been involved in the development of a flywheel energy storage system for possible spacecraft applications. Flywheels have potential advantages over traditionally used battery storage systems. These include

- high power density
- long life
- deep depth of discharge
- broad operating temperature range

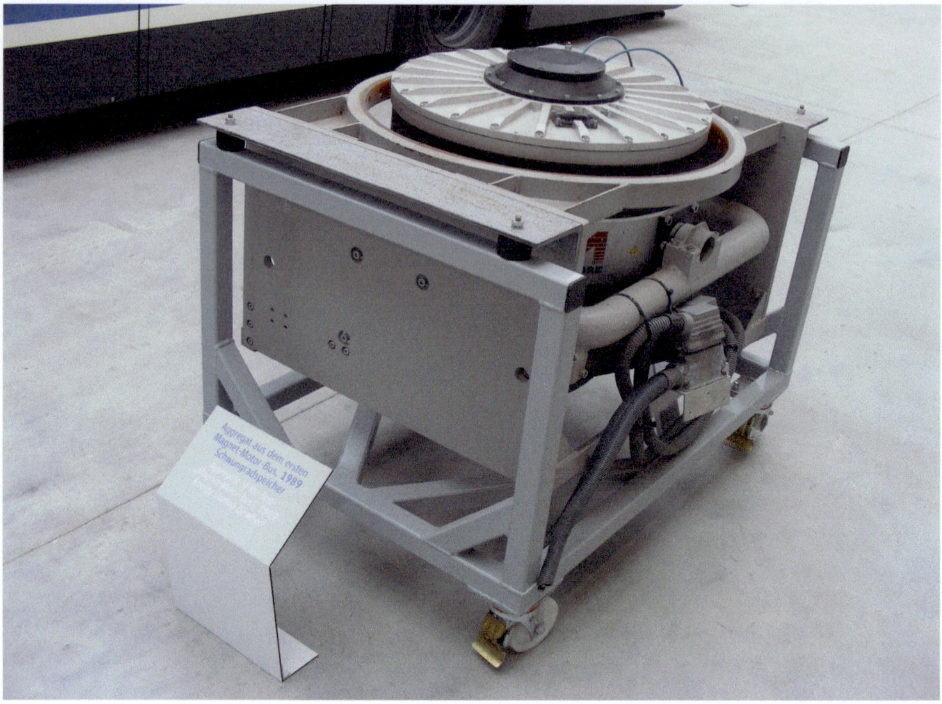

Fig. 3.12 Commercial flywheel energy storage module built in 1989 and once installed in a Munich city bus. User Mattes (2009) CC0 1.0 Public domain. https://creativecommons.org/publicdomain/zero/1.0/deed.en

A diagram of a NASA prototype unit (named G2) is shown in Fig. 3.14. The flywheel consists of a carbon fibre rim attached to a titanium hub and rotates at a maximum of 60,000 rpm. This unit stores a total of 525 Wh of energy and has an output of 1 kW, for a duration of just over half an hour. A photograph of the prototype unit is shown in Fig. 3.15.

3.3.3.5 Nuclear Fusion Experiments

Many nuclear fusion experiments use very large amounts of power for relatively short periods of time. Flywheels are an ideal means of storing electrical energy for these short power bursts. The Joint European Torus (or JET) is an example of a facility that used flywheel energy storage for fusion research. The Joint European Torus was a tokamak-based magnetic fusion reactor design located at the Culham Centre for Fusion Energy in the United Kingdom that operated from 1983 to 2023. Typical experimental runs (or "shots") lasted about 20 s and could use up to 1 GW of peak electrical power. Since this is more than can be supplied by the electric grid, energy was stored in two flywheels that were used to provide electricity during the experiment. Each flywheel had a 9 m

Fig. 3.13 Audi R18 e-tron quattro uses regenerative braking to charge a flywheel to provide additional power during acceleration. Morio (2012) CC BY-SA 3.0. https://creativecommons.org/licenses/by-sa/3.0/deed.en

diameter rotor and weighed over 700 tonnes. At full speed the rotors turned at 225 rpm and they were cycled down to about half speed during the 20 s discharge. Each flywheel provided up to about 500 MW output. Recharging the flywheels took about 9 min and was accomplished by two 8.8 MW (~ 12,000 hp) motors, one for each rotor. This represents an ideal application of flywheel energy storage where high power and reasonably rapid cycle time is needed.

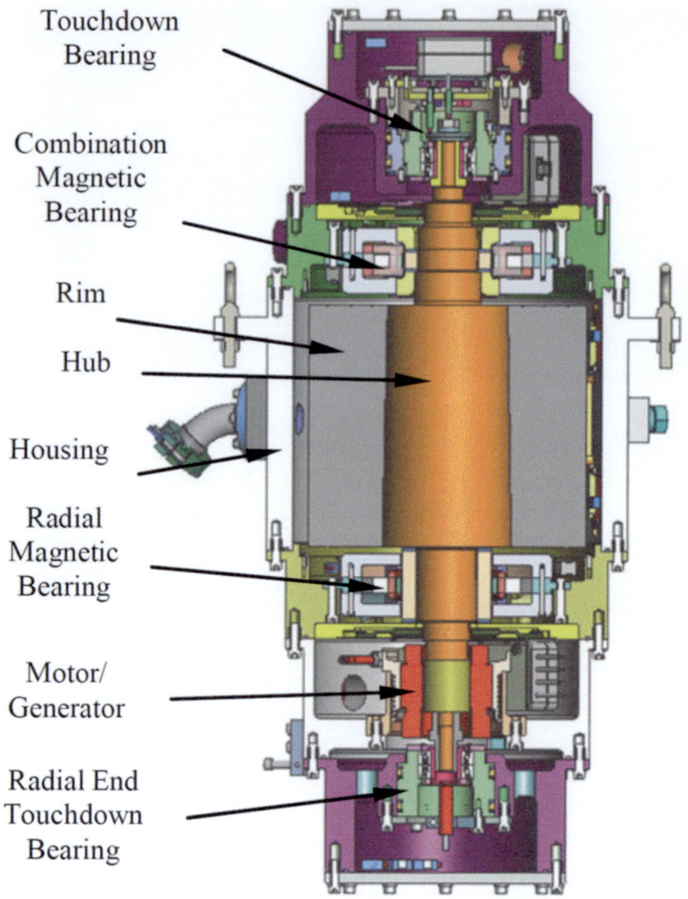

Fig. 3.14 Diagram of NASA's G2 flywheel energy storage module. Figure 1 from Nagorny et al. (2007) NASA Public domain

Fig. 3.15 Photograph of NASA's G2 flywheel energy storage module. NASA (2005) Public domain

References

Amber Kinetics (2024) The A32. https://amberkinetics.com/

Callio (2024) Pumped hydro energy storage. https://callio.info/energy-storage/opportunities/

Cava F, Kelly J, Peitzke W, Brown M, Sullivan S (2016) Advanced rail energy storage: green energy storage for green energy, chap 4. In: Letcher TM (ed) Storing energy—with special reference to renewable energy sources. Elsevier, Amsterdam, pp 69–86. https://doi.org/10.1016/B978-0-12-803440-8.00004-X

Dace A (2010) English: Hick Hargreaves No 303 steam engine, near to Forncett St Mary, Norfolk, Great Britain. https://commons.wikimedia.org/wiki/File:Hick_Hargreaves_No_303_Steam_Engine_-_geograph.org.uk_-_1991615.jpg

Energy Vault (2024) Energy storage built to your needs. https://www.energyvault.com/

Enqvist T (2006) EMMA: a new underground cosmic-ray experiment. J Phys Conf Ser 39:478–480. https://doi.org/10.1088/1742-6596/39/1/125

Galant S, Peirano E, Debarberis L (2013) Electricity storage: a new flexibility option for future power systems Power Syst 69:247−284. https://doi.org/10.1007/978-1-4471-4549-3_7

Gravitricity (2024) Gravitricity—long-life, distributed, underground energy storage. https://gravitricity.com/

Gravity Power LLC (nd) Low-cost energy storage with minimal environmental impact. https://www.gravitypower.net/

Keimzelle (2022) Energy Vault testing tower in Castione-Arbedo (January 2022). https://commons.wikimedia.org/wiki/File:Energy_Vault_Castione_(2022-01-26).jpg

Komori M, Kato H, Asami K (2022) Suspension-type of flywheel energy storage system using high T_C superconducting magnetic bearing (SMB). Actuators 11:215. https://doi.org/10.3390/act11080215

London Midland (2008) Parry people movers for Stourbridge branch line London Midland (03 January 2008). https://web.archive.org/web/20080517110918/http://www.londonmidland.com/index.php/news/news_items/view/23

Menendez J, Schmidt F, Loredo J (2020) Comparing subsurface energy storage systems: underground pumped storage hydropower, compressed air energy storage and suspended weight gravity energy storage. E3S Web Conf 162:01001 https://doi.org/10.1051/e3sconf/202016201001

Morio (2012) Audi R18 e-tron quattro no1 top view. https://commons.wikimedia.org/wiki/File:Audi_R18_e-tron_quattro_no1_top_view_2012_WEC_Fuji.jpg

Nagorny A, Jansen R, Kankam M (2007) Experimental performance evaluation of a high speed permanent magnet synchronous motor and drive for a flywheel application at different frequencies NASA/TM-2007-214428. https://ntrs.nasa.gov/api/citations/20080008840/downloads/20080008840.pdf

NASA (2005) G2 flywheel module. https://commons.wikimedia.org/wiki/File:G2_front2.jpg

Pjrensburg (2012) This image shows the main components of a typical cylindrical flywheel rotor assembly. https://commons.wikimedia.org/wiki/File:Example_of_cylindrical_flywheel_rotor_assembly.png

Power Technology (2021) Beacon power Stephentown—flywheel energy storage system power technology (28 August 2021). https://www.power-technology.com/marketdata/beacon-power-stephentown-flywheel-energy-storage-system-us/?cf-view

Rudge S (2024) Europe's deepest mine to become colossal gravity battery. Min Miner Today (13 March 2024). https://m-mtoday.com/news/europes-deepest-mine-to-become-colossal-gravity-battery/

Sarffman (2013) Measurement station of EMMA experiment in Pyhäsalmi mine. https://commons.wikimedia.org/wiki/File:Emma_koe.JPG

Skinner M, Mertiny P (2022) Energy storage flywheel rotors—mechanical design. Encyclopedia 2:301–324. https://doi.org/10.3390/encyclopedia2010019

Staff Writer (2024) European deep mine operators looking into underground energy storage. Mining.com (14 May 2024). https://www.mining.com/european-deep-mine-operators-looking-into-underground-energy-storage/

Tisheva P (2024) Gravitricity plans 2-MW gravity energy store at Finnish mine. Renew Now (05 February 2024). https://renewablesnow.com/news/gravitricity-plans-2-mw-gravity-energy-store-at-finnish-mine-847637/

User Mattes (2009) Schwungradspeicher, Hersteller: magnet motor, Bj. 1989, wurde einst in einen Linienbus der Stadtwerke München verbaut. https://commons.wikimedia.org/wiki/File:Schwungradspeicher_(Magnet_Motor),_Bj_1989_-_Exponat_im_MVG_Museum.jpg
Volkov V (2008) Gyrobus G3. https://commons.wikimedia.org/wiki/File:Gyrobus_G3-1.jpg

Sensible Heat Energy Storage

4

4.1 Introduction

Energy may be stored by using the thermal properties of a material. These techniques can be instituted on a wide variety of scales, from individual residential energy storage to storage on a city or regional scale. Three different material properties allow for the storage of thermal energy,

- sensible heat
- latent heat
- thermochemical reactions

Sensible heat is related to the change in temperature of a material to which heat is applied that results from its heat capacity and is discussed in the present chapter. Latent heat and thermochemical energy storage technologies are discussed in Chap. 6.

4.2 Basics of Sensible Heat Energy Storage

The sensible heat associated with a material is the energy that is needed to heat the material from a lower temperature to a higher temperature when there are no phase transitions between the two temperatures. It is also the energy that can be extracted from the material by cooling it from the higher temperature to the lower temperature. The relationship between the thermal energy, Q, and the change in temperature, ΔT, is

$$Q = C\Delta T. \tag{4.1}$$

R. A. Dunlap, *Renewable Energy Storage*, Synthesis Lectures on Renewable Energy Technologies, https://doi.org/10.1007/978-3-031-88942-4_4

where C is the heat capacity of the material. When Q is in Joules and temperature is in Celsius (or Kelvin) then the heat capacity is in units of J/°C. When selecting materials for a particular application it is generally reasonable to consider the intrinsic properties of the material, that is the heat capacity for a particular quantity of the material. This may be expressed as the heat capacity per unit mass, c, (called the *specific heat capacity*, or more commonly, just the *specific heat*), the heat capacity per mole, C_{mol}, (called the *molar heat capacity*) or the heat capacity per unit volume, C_{vol}, (called the *volumetric heat capacity*). The relationship of these quantities to the heat capacity is given by

$$c = \frac{C}{m}, \tag{4.2}$$

where m is the sample mass,

$$C_{mol} = \frac{CM}{m}, \tag{4.3}$$

where M is the molecular weight of the material and

$$C_{vol} = \frac{C}{V} = \frac{c}{\rho}, \tag{4.4}$$

where V is the sample volume and ρ is the material density. The volumetric heat capacity is generally only used for liquids and solids but not for gases.

4.3 Sensible Heat Energy Storage Materials

For particular heat storage applications, the choice of a storage material must consider the thermal properties of that material. Some relevant properties are

- the phase of the material, i.e. solid or liquid, may be of relevance to the way in which the stored heat will be distributed and used
- the volumetric heat capacity is of relevance if the space for the storage facility is limited
- the specific heat is important if the mass of the storage facility is limited
- the range of temperatures over which the material can be used (i.e. temperatures for which it is stable and does not undergo any phase transitions)

Other factors such as toxicity and cost should also be considered. Table 4.1 gives the relevant properties of some common materials that are used for thermal energy storage. Clearly, from an energy standpoint, water stands out as a suitable material. Applications of sensible heat energy storage are discussed in the remainder of the chapter.

Table 4.1 Typical values for thermal properties of some common materials for sensible heat energy storage

Material	Specific heat [J/(kg·°C)]	Density (kg/m^3)	Volumetric heat capacity [kJ/(m^3·°C)]
Water	4186	1000	4186
Wood (pine)	2800	500	1400
Stone (solid)	879	2560	2250
Stone (loose)	879	1500	1320
Sand	816	1600	1306
Concrete	653	2300	1502
Iron	460	7855	3613

4.4 Residential Heat Storage

Perhaps the simplest and most obvious application for sensible heat energy storage is for residential use. Residences which utilize solar panels for space heating or domestic hot water heating purposes require diurnal heat storage to provide thermal energy during the night or during periods of low solar insolation. Water is an appropriate storage material, not only from a thermal properties standpoint, but because it is a common non-toxic liquid that is suitable for distributing heat through a central heating system or for providing domestic hot water.

There are several different approaches to the use of stored energy from solar collectors. Figure 4.1 shows a simple passive system for heating water for domestic hot water use. Figure 4.2 shows a photograph of this type of installation. The storage tank is mounted on the roof immediately adjacent to the solar collector. Pressurized cold water runs through the tank and is heated and returned as hot water for residential use. The system shown in Fig. 4.1 uses an auxiliary natural gas fired heater to maintain appropriate water temperature. Such systems may not be suitable for colder climates because of heat loss from the external storage tank.

An active solar heating system, i.e., one in which water or a working fluid is pumped through a solar collector and storage tank, is illustrated in Fig. 4.3. In this case a ground level or basement storage tank can be used, as shown in the figure. This arrangement has the potential of utilizing an internal storage tank and thereby reducing the heat loss that is characteristic of an external tank. Figure 4.4 shows a schematic of a solar space heating system using a heat exchanger in the storage tank. Figure 4.5 shows a photograph of a typical basement installation of a solar heated water storage tank for space heating purposes.

The design of a heat storage system as illustrated in Figs. 4.3, 4.4 and 4.5 for space heating purposes requires a careful consideration of the area of the solar collectors to

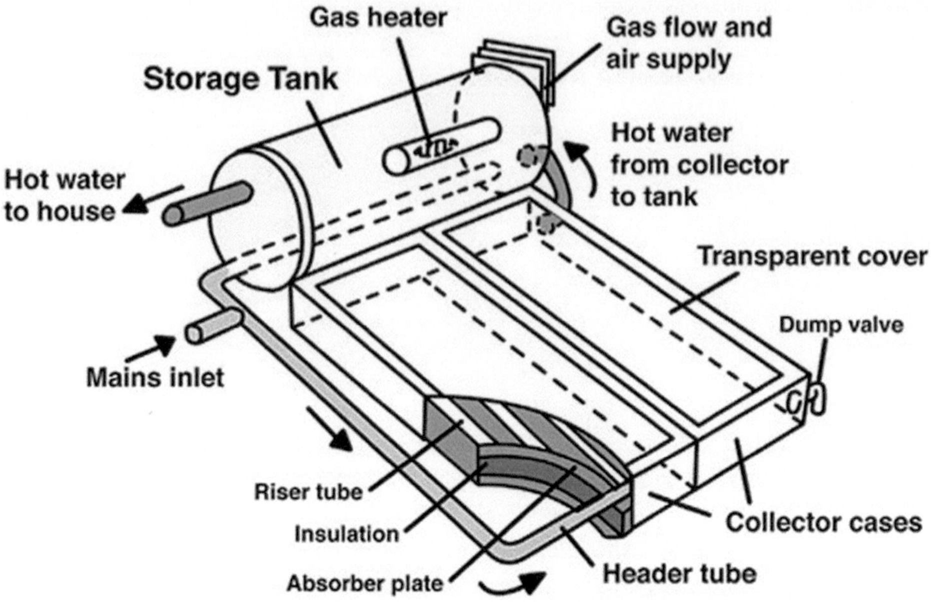

Fig. 4.1 Diagram of a solar thermal domestic hot water system using a flat plate collector and a roof mounted storage tank. Riedy et al. (2020) Copyright Commonwealth of Australia CC BY 4.0. https://creativecommons.org/licenses/by/4.0/

ensure sufficient energy to satisfy heating requirements. It is also necessary to consider the size of the storage tank in order to provide adequate heat during the night and during days of low insolation. These factors obviously relate to the local climate as well as details of the building construction.

As an example, we can estimate the winter diurnal heating requirement for a typical house with a volume of 700 m^3. A reasonably well insulated house of this size might require 2.8 kJ/m^3 per hour per °C temperature difference between the inside of the house and the outside. We consider an average daily outside temperature of − 6 °C (typical of a cold winter day in Boston, MA) and require an inside temperature of 19 °C. A simple calculation gives the daily heating requirement as

$$\left(2800\frac{\mathrm{J}}{\mathrm{m}^3\,{}^\circ\mathrm{C\,h}}\right)\left(24\frac{h}{d}\right)(700\,\mathrm{m}^3)(25\,{}^\circ\mathrm{C}) = 1.18\times10^9\frac{\mathrm{J}}{\mathrm{d}}. \tag{4.5}$$

To determine the size of a suitable storage system as shown in Fig. 4.6, we combine Eqs. (4.1) and (4.2) and solve for mass,

$$m = \frac{Q}{c\Delta T}. \tag{4.6}$$

Fig. 4.2 A roof mounted solar energy storage tank coupled to flat plate solar collectors. Surek (2006) CC BY-SA 3.0. https://creativecommons.org/licenses/by-sa/3.0/deed.en

The temperature difference, ΔT, in this equation gives the difference in the temperature of the storage tank when it is at a maximum after being heated from the solar collector, T_{max}, and when it is at a minimum after being used to heat the house overnight, T_{min}. Using an estimate of these temperatures as $T_{max} = 85$ °C and $T_{min} = 35$ °C we find the mass of water needed to store the energy given in Eq. (4.6) as

$$m = \frac{1.18 \times 10^9\,\text{J}}{4186\frac{\text{J}}{\text{kg}\times{}^\circ\text{C}} \times (85 - 35^\circ\text{C})} = 5600\,\text{kg}. \tag{4.7}$$

This mass represents a volume of water of 5.6 m^3. This can be stored in a cylindrical tank 2 m high by about 2 m in diameter.

On the basis of Table 4.1 it is possible to consider materials other than water for residential heat storage. From the materials in the table, iron is seen to be a close second to water in terms of volumetric heat capacity. Although one does not think of iron as a particularly expensive material, a calculation of the required mass of iron to replace water for the calculation in Eq. (4.7) yields a value of around 50,000 kg, which gives an unreasonable cost, even for this modestly priced material. It is seen from the table

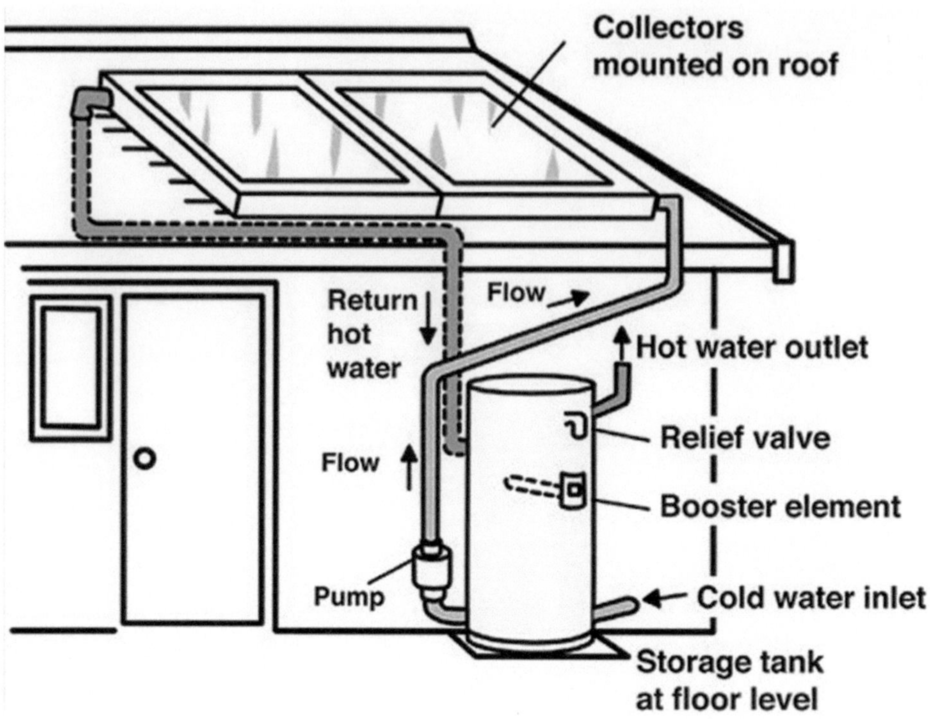

Fig. 4.3 Active solar heating system utilizing roof mounted flat plate collectors and a ground level storage tank. Riedy et al. (2020) Copyright Commonwealth of Australia CC BY 4.0. https://creativecommons.org/licenses/by/4.0/

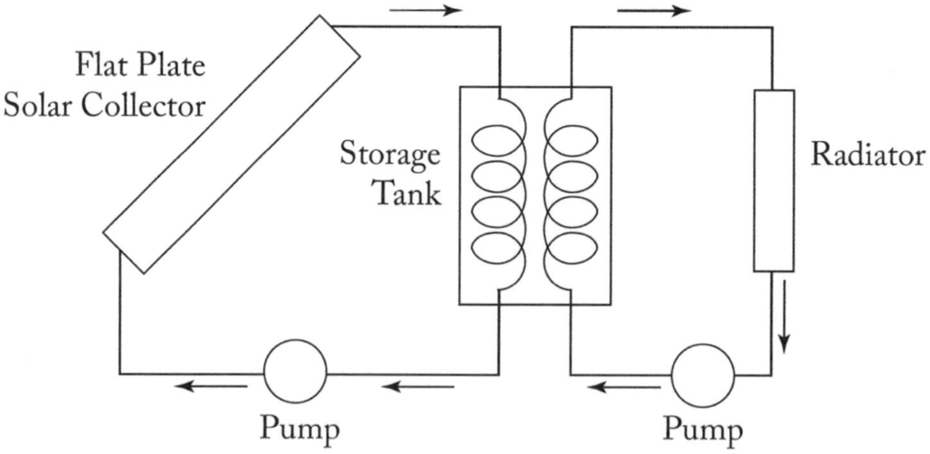

Fig. 4.4 Simple system for storing thermal energy from a flat plate solar collector for residential use during the night

Fig. 4.5 Photograph of interior thermal storage tank. PommeDe Therme (2024) CC0 1.0 Public domain. https://creativecommons.org/publicdomain/zero/1.0/deed.en

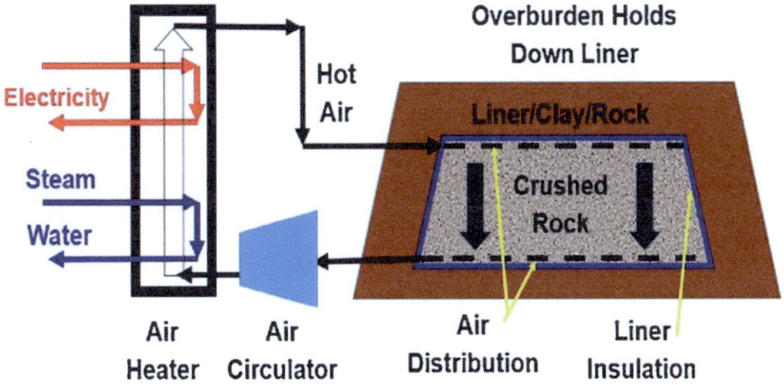

Fig. 4.6 Residential thermal storage using the heat capacity of rocks. Flow directions are shown for the charge mode. Figure 6 Reprinted from Forsberg et al. (2018) Copyright (2018) with permission from Elsevier CC BY 4.0. https://creativecommons.org/licenses/by/4.0/

that stone may be the most reasonable alternative to water. The specific heat is roughly twice that of iron and it is, as a naturally occurring material, readily and inexpensively available. A typical system utilizing loose stone for residential heat storage is illustrated in Fig. 4.6. Although the volumetric heat capacity is only a little more than half that of water, such a system may have advantages in some situations. As the figure shows, loose packed material, specifically rock, can be used efficiently as a heat storage medium for hot air space heating systems. This provides a simpler alternative to water-based heat storage for hot air heating systems, where a water-to-air heat exchanger is needed. The use of rocks for heat storage, combined with air as the working fluid to transport heat, does not limit the operating temperature to less than 100 °C, as is the case for water-based systems. While this is not generally a consideration for residential system incorporating flat plate solar collectors, it is a factor for concentrating solar collector systems that are commonly used for grid integrated systems as discussed below. A number of studies have been undertaken to evaluate parameters such as rock size and packing fraction, along with air flow rate, on system performance (e.g. Choudhury et al. 1995; Singh et al. 2013).

4.5 District Heating Systems

While the system described above is suitable for an individual residence, the use of thermal storage for an entire community, i.e. housing development, town, city, etc. requires a system with a much greater heat storage capacity. It would also require a greater capacity to harvest heat to be stored. A generic district heating system is illustrated in Fig. 4.7. The primary heat source can be fossil fuel, as it has been in the past for many such systems, or a renewable source, such as geothermal or solar, as discussed in the examples below. In fact, the development of district heating systems over the years, from fossil fuel based to renewable energy based is summarized in Fig. 4.8. Here, four clearly defined generations of district heating systems are illustrated with appropriate years of use indicated. The diagram shows the importance of heat storage in all district heating systems and the introduction of long-term seasonal storage in the most recent generation. The 5th generation system, as shown in Fig. 4.9, will incorporate both heat storage and cold storage to accommodate both winter heating requirements and summer air conditioning requirements, respectively.

A major component of any renewable energy-based district heating (and/or cooling) system is thermal energy storage. Thermal energy storage can be short-term, to deal with daily fluctuations of available energy, such as solar. It can also be long-term to deal with seasonal fluctuations of available energy. Figure 4.10 shows a breakdown of the different types of thermal energy storage that could contribute to a renewable district heating (or cooling) system. In the present chapter we look at short-term and long-term options using sensible heat. In the next chapter latent heat energy storage technologies will be considered.

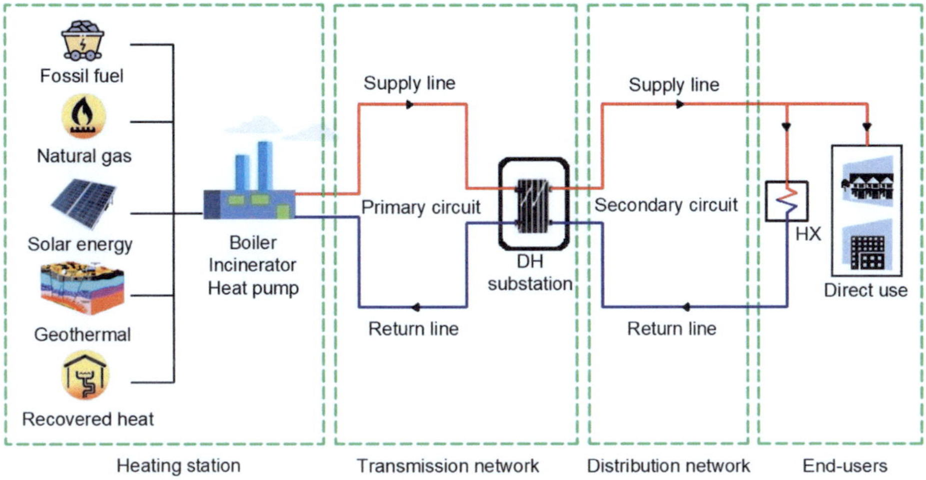

Fig. 4.7 Diagram showing the basic components of a district heating system. Figure 1 from Sarbu et al. (2022) Copyright 2022 by the Authors CC BY 4.0. https://creativecommons.org/licenses/by/4.0/

In the remainder of this section, three community-based renewable energy heating systems utilizing energy storage are described. These are the geothermal energy district heating system of Reykjavik, Iceland, the solar energy heat storage system for the Drake Landing Solar Community in Okotoks, Alberta, Canada and the solar heating system in Brædstrup, Denmark.

4.5.1 Reykjavik District Heating System

Geothermal energy can be utilized as a source of electricity as well as a source of thermal energy. Direct geothermal energy is utilized by a number of major cities worldwide who fulfill a portion of their heating needs from this source. These cities include Beijing, Bucharest, Budapest, Paris, Rome and Sophia.

Iceland has abundant geothermal resources. These resources provide about 30% of the electricity used in the country and nearly 90% of the country's 382,000 residents obtain space heating from geothermal resources. There are 29 separate district heating systems in Iceland that provide heat to various communities. In the capital city, Reykjavik, 95% of the buildings are connected to a centralized district heating system which utilizes geothermal energy, and this system is described below.

The Reykjavik district heating system utilizes heat from four geothermal fields. One of these fields, Nesjavellir, as shown in Fig. 4.11, is a high temperature geothermal resource while the other three fields, as summarized in Table 4.2, are low temperature resources.

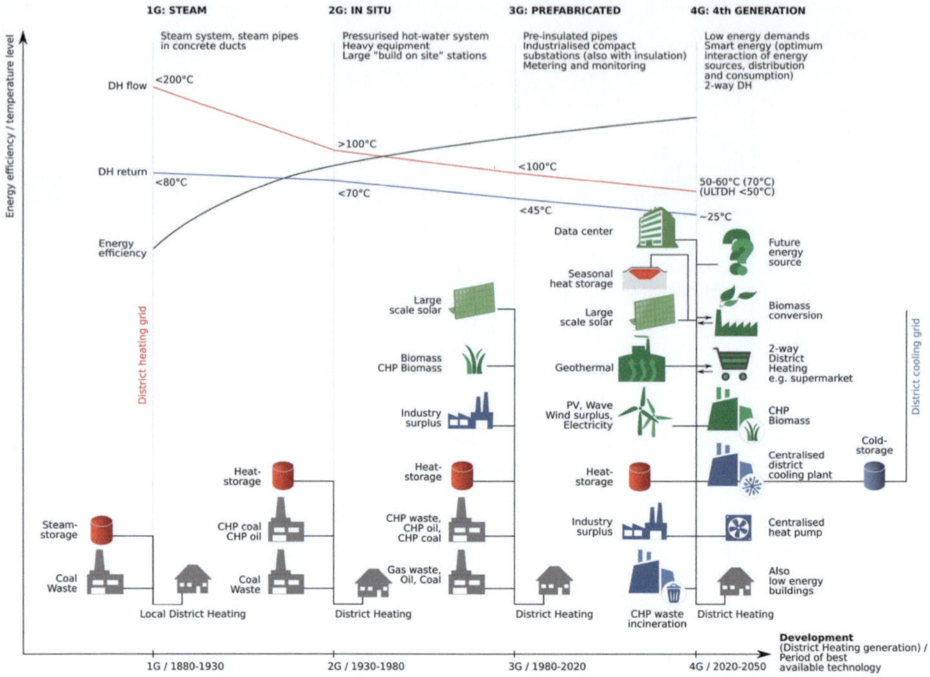

Fig. 4.8 Four generations of district heating systems showing the progression of energy sources. Based on Lund et al. (2014). Mrmw and Andol (2018) CC0 1.0 Public domain. https://creativecommons.org/publicdomain/zero/1.0/deed.en

Figure 4.12 shows a simplified diagram of the Nesjavellir plant and its connection to the Reykjavik heating system.

The geothermal water is carried from the geothermal fields to Reykjavik through pipelines as illustrated in Fig. 4.13. The interior of Orkuveita Reykjavikur, the municipal utility serving Reykjavik with electricity and heat, is shown in Fig. 4.14. From this facility, geothermally heated water is distributed to end users. In addition to providing space heating for buildings in Reykjavik, geothermal resources are used to heat roadways to reduce ice during the winter months (see Fig. 4.15).

Storage of geothermal heat is an important component of the Reykjavik district heating system. The use of geothermal water from Reykir is an example of how storage facilities are important for the distribution of heat to the community. The geothermal water from Reykir flows to a storage facility consisting of six tanks outside of Reykjavik which have a total capacity of 5.4×10^7 L. From there, the water flows to the city where it fills six additional storage tanks with a capacity of 2.4×10^7 L. This latter facility, known as the Perlan (Icelandic for "The Pearl"), sits atop Öskjuhlíð, a hill in central Reykjavik, which is 61 m above sea level. Figure 4.16 shows a photograph of the Perlan. The central dome

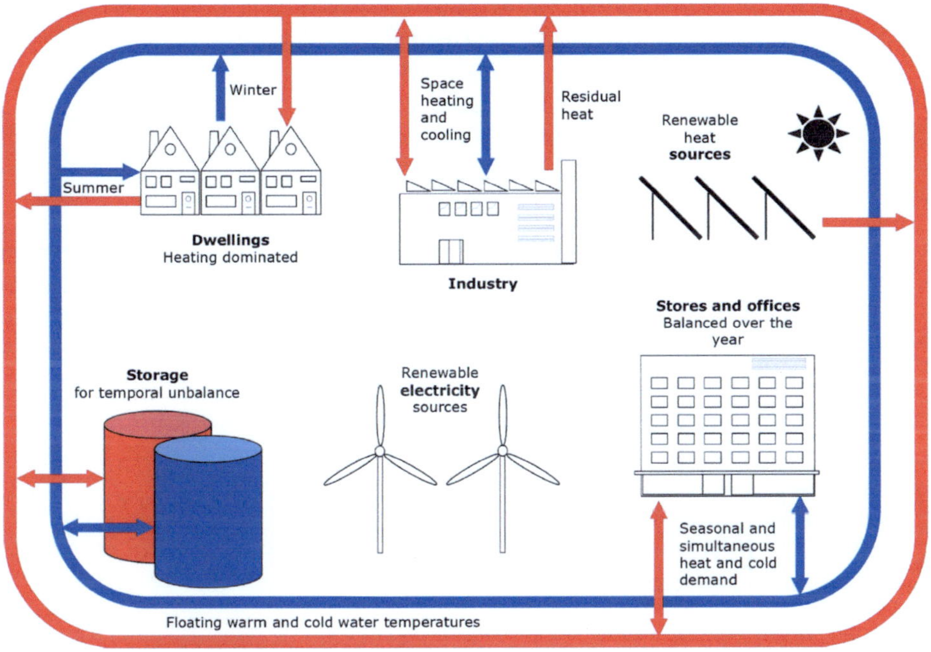

Fig. 4.9 Schematic diagram of a generic 5th generation district heating and cooling system. Figure 1 from Boesten et al. (2019) Copyright the Authors CC BY 4.0. https://creativecommons.org/licenses/by/4.0/

above the storage tanks houses a rotating restaurant, a planetarium and a gift shop. From the Perlan, geothermal water is distributed to the city through nine pumping stations. Because district heating requirements are not constant in time, the storage tanks provide a buffer for periods of high demand.

4.5.2 Drake Landing Solar Community

The Drake Landing Solar Community in Okotoks, Alberta, Canada is a community consisting of 52 single family homes as shown in the photograph in Fig. 4.17. The community was constructed in 2005 and the homes range in size from 139 to 155 m^2 in floor area. The community is designed to obtain nearly all its heating from solar energy. The goal of the design is to supply virtually all the heat required during the winter months, when the heating requirements are the greatest and the insolation is the least, using thermal energy that is stored during the summer months when heating requirements are minimal. Domestic hot water is also supplemented with solar energy.

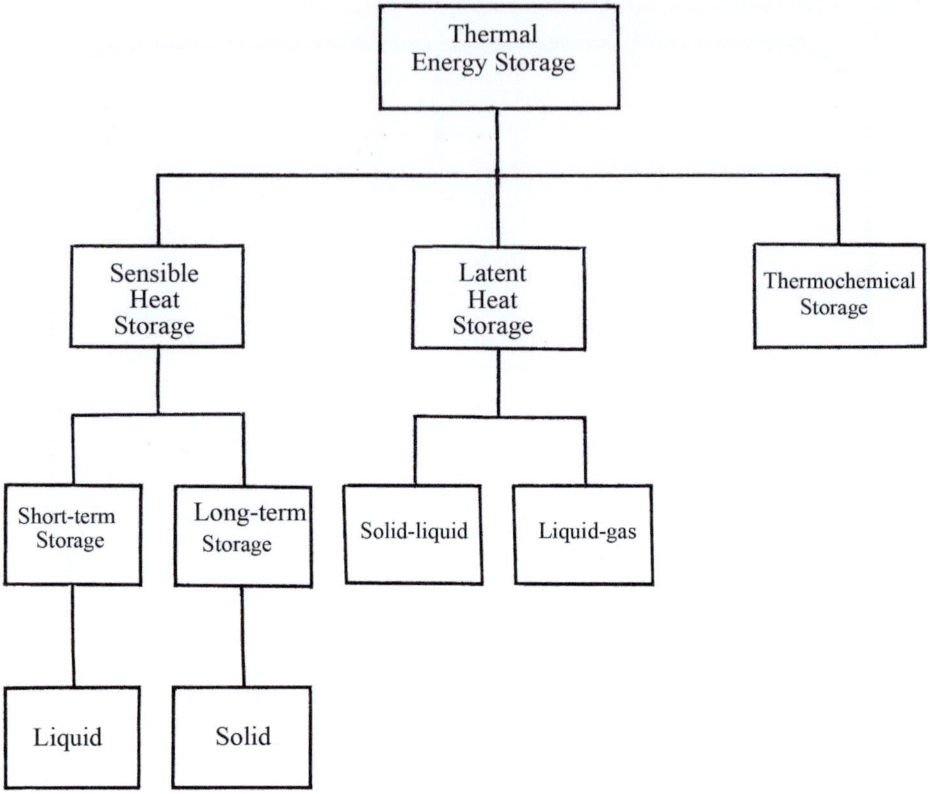

Fig. 4.10 Classification of district heating thermal energy storage methods

Fig. 4.11 Nesjavellir geothermal power plant located about 30 km east of Reykjavik. Krapf (2008) CC BY-SA 3.0. https://creativecommons.org/licenses/by-sa/3.0/deed.en

Table 4.2 Low temperature geothermal fields that contribute to the Reykjavik district heating system

Field	Temperature (°C)	Flow (kg/s)
Reykir	85–95	1700
Laugarnes	125–130	330
Elliðaár	85–95	220

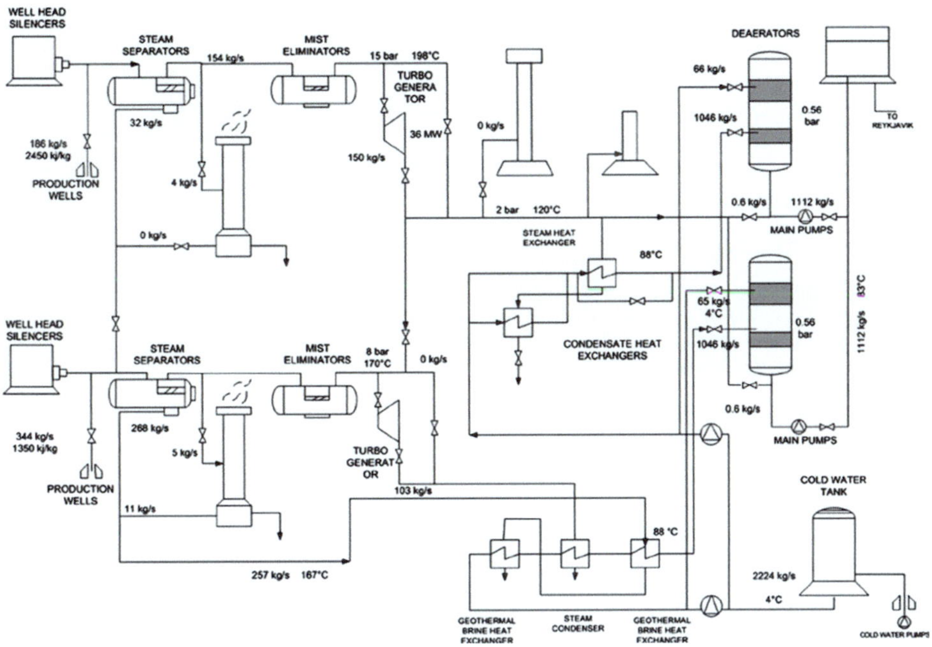

Fig. 4.12 Diagram of the Nesjavellir plant which supplies electricity and district heating to Reykjavik, Iceland. Adapted from Ballzus et al. (2000). Figure 22 Reprinted from Rubio-Maya et al. (2015) Copyright (2015) with permission from Elsevier

The community heating system contains 798 single glazed flat plate solar collectors mounted on 45 m^2 detached garages as seen in the figure. The garages are interconnected to one another with breezeways in order to provide a contiguous platform for mounting the solar collectors. In the summer, these collectors provide an average of about 1.5 MW of thermal power during the day. Fluid (water with non-toxic glycol added) is circulated through the solar collectors, where it is heated, and then into a series of 144 boreholes, each about 35 m deep, where the heat is transferred to the soil. Electricity for the pumps required to circulate the fluid is obtained from photovoltaic collectors to ensure net carbon-neutrality for the system.

By the end of the summer, the soil in the vicinity of the boreholes reaches a temperature of about 80 °C. In the winter. when heat is required in the homes, cold water (with

Fig. 4.13 Pipeline from Nesjavellir Geothermal Power Plant to Reykjavik. Bowlingbal (2009) CC BY-SA 3.0. https://creativecommons.org/licenses/by-sa/3.0/deed.en

glycol) is circulated through the boreholes where it is heated. It is then circulated through heat exchangers in the homes to provide space heating as well as domestic hot water. Figure 4.18 shows the distribution of energy sources that deliver heat to the community throughout the year. Since the community was established, more than 90% of the required heat has been provided by solar energy, as well as about 50% of the domestic hot water. The remaining heat and hot water are provided by the combustion of natural gas, shown as "boiler" in the figure. Due to aging of system components, the solar heating system at Drake Landing is scheduled for decommissioning by the end of 2025 (Strand 2024). At that time, solar heating will be replaced by other technologies.

4.5.3 Brædstrup Fjernvarme

The Brædstrup Fjernvarme (Fjernvarme = Danish for heating) system in the town of Brædstrup, Denmark (population about 3600) uses flat plate solar collectors together with both short-term and long-term thermal storage. The system was developed in two stages. Stage I, which utilized 8000 m^2 of solar collectors, became operational in 2007. Stage II, which expanded the solar collector area to 18,600 m^2, became operational in 2011. Figure 4.19 shows some of the solar collectors in the Brædstrup Fjernvarme system.

Fig. 4.14 Interior of Orkuveita Reykjavikur, the municipal utility serving Reykjavik with electricity and heat. Gipe (2011) CC BY-SA 4.0. https://creativecommons.org/licenses/by-sa/4.0/deed.en

The Brædstrup is a hybrid solar-natural gas system. The solar collectors provide thermal energy while the natural gas generating station is a combined heat and power (CHP) facility that produces both electricity and heat. The electricity produced by the natural gas generating station is fed into the grid. The heat produced from the natural gas fired station is combined with heat from the solar collectors and is stored in two above ground thermal water storage tanks. Long-term (i.e. seasonal) heat storage is accomplished, as for the Drake Landing Community, using soil heated by circulating hot water through bore holes. In the winter, heat stored during the summer is recovered by circulating cold water through the bore holes.

At present, the solar contribution to the total heat supplied is about 20%, while the remaining 80% comes from the thermal output of the natural gas CHP plant.

Fig. 4.15 Pipes that carry geothermally heated water underneath roadway. Image shows road under construction. Nygaard (2009) CC BY 2.0. https://creativecommons.org/licenses/by/2.0/

Fig. 4.16 The Perlan geothermal storage tanks in Reykjavik, Iceland (image Richard A. Dunlap)

Fig. 4.17 Aerial photograph of the Drake Landing Solar Community. Figure 1 Reprinted from Sibbitt et al. (2012) Copyright (2012) with permission from Elsevier. CC BY-NC-ND 3.0. https://creativecommons.org/licenses/by-nc-nd/3.0/deed.en

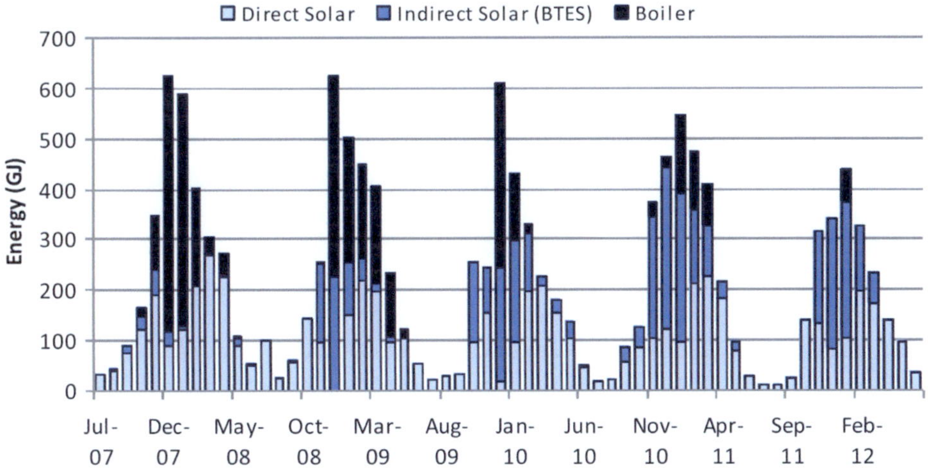

Fig. 4.18 Monthly energy supplied to the distribution loop at Drake Landing Solar Community from 2007 to 2012. Figure 5 Reprinted from Sibbitt et al. (2012) Copyright (2012) with permission from Elsevier. CC BY-NC-ND 3.0. https://creativecommons.org/licenses/by-nc-nd/3.0/deed.en

Fig. 4.19 Solar collectors at Brædstrup Fjernvarme energy facility. DKFjernvarmen (2011) CC BY-SA 4.0. https://creativecommons.org/licenses/by-sa/4.0/deed.en

4.6 Grid Integrated Systems

The most notable examples of grid-integrated generating facilities that utilize thermal energy storage are systems based on concentrating solar collectors. These typically use the sensible heat associated with molten salt to store thermal energy. Figure 4.20 shows a typical concentrating central receiver system with molten salt thermal storage. The same concept can also be applied to parabolic collector systems. "Cold" molten salt is stored in the cold salt storage tank. It is pumped through the focus area of the collector where it is heated and is then pumped to the hot salt storage tank. To generate electricity, the hot molten salt is pumped through a heat exchanger to heat water and produce steam. This steam is then used to drive a turbine/generator to produce electricity. The molten salt is then returned to the cold storage tank. This thermal storage approach has also been used in conjunction with parabolic trough solar collectors and the basic design for such a system is illustrated in Fig. 4.21. Salt storage tanks at a typical parabolic trough solar generating station are shown in Fig. 4.22.

Fig. 4.20 The 50-megawatt solar thermal power station in Hami, Xinjiang Uygur autonomous region. Csp.guru (2022) CC BY-SA 4.0. https://creativecommons.org/licenses/by-sa/4.0/deed.en

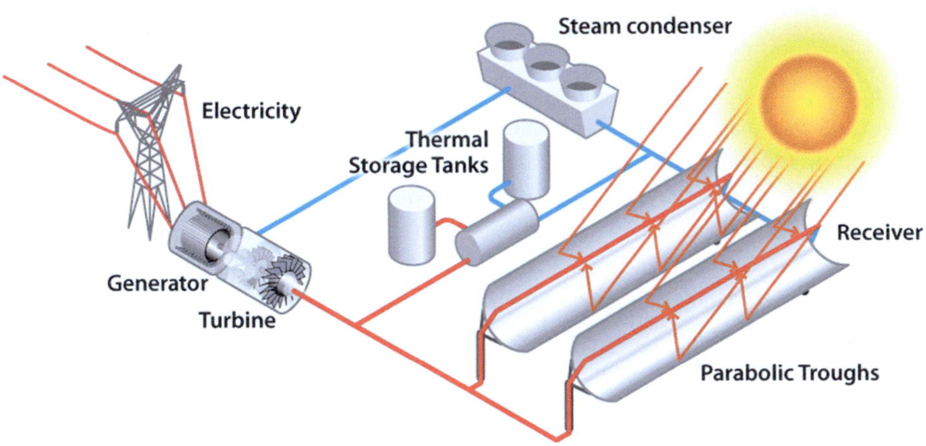

Fig. 4.21 Diagram of a concentrating solar collector showing thermal energy storage in molten salt tanks. U. S. Department of Energy (2016) Public domain

Fig. 4.22 Molten salt storage tanks at the Solana Generating Station, a parabolic trough solar facility near Gila Bend, Arizona. Photo by Dennis Schroeder. ENERGY.GOV (2012) Public domain

The salt that is most commonly used for systems as described above, is a eutectic mixture of 60% KNO_3 and 40% $NaNO_3$. Some relevant properties of this salt mixture are given in Table 4.3. The cold salt storage tank stores the molten salt at a temperature of around 285 °C and the hot salt storage tank stores the heated salt at a temperature of around 565 °C. Well insulated storage tanks can maintain the salt temperature for about a week and the estimated efficiency of thermal storage is around 99%, meaning that the electricity generated after thermal storage of the salt is 99% of that which would have been generated immediately. The size of the storage tanks depends on the required power output and duration. As an example, two tanks (one cold and one hot) with dimensions of 24.4 m in diameter by 9.1 m high, could provide an electrical output of 50 MW for a period of 8 h. In the summer of 2013, the Gemasolar Thermosolar central receiver power plant in Fuentes de Andalucía, Seville, Spain was the first such facility to generate power continuously for 24 h a day over an extended period of time (36 days) using thermal energy stored during the day to provide electrical output during the night.

Table 4.3 Some relevant properties of a eutectic mixture of 60% KNO_3 and 40% $NaNO_3$ salt compared with water and loose rock

Property	Salt	Water	Loose rock	Units
Melting point	240	0	–	°C
Density	2200	1000	1500	kg/m^3
Specific heat	1530	4186	879	J/(kg·°C)
Volumetric heat capacity	3366	4186	1320	$kJ/(m^3\cdot°C)$

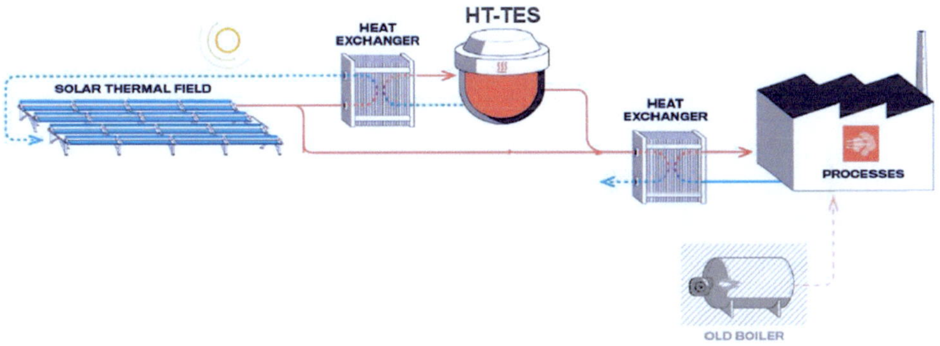

Fig. 4.23 Schematic of concentrating solar system utilizing a rock bed (HT-TES) for thermal storage. Figure 1 Reprinted from Muhammad et al. (2023) Copyright (2023) with permission from Elsevier. CC BY 4.0. https://creativecommons.org/licenses/by/4.0/

It has also been proposed that rock beds could be used for thermal storage for concentrating solar collector systems. A possible system for this purpose is illustrated in Fig. 4.23. Air is heated by concentrated solar radiation in the central receiver. This hot air this then circulated through a bed of loose rock. The hot rock stores the thermal energy, and this energy is recovered as needed and transferred to water in a heat exchanger to produce steam to run turbines. Figure 4.24 shows a possible design of a rock bed thermal energy storage unit. A comparison of the properties of salt, water and rock is shown in Table 4.3. While rock has a lower heat capacity than water or salt, it is useable over a greater range of temperatures.

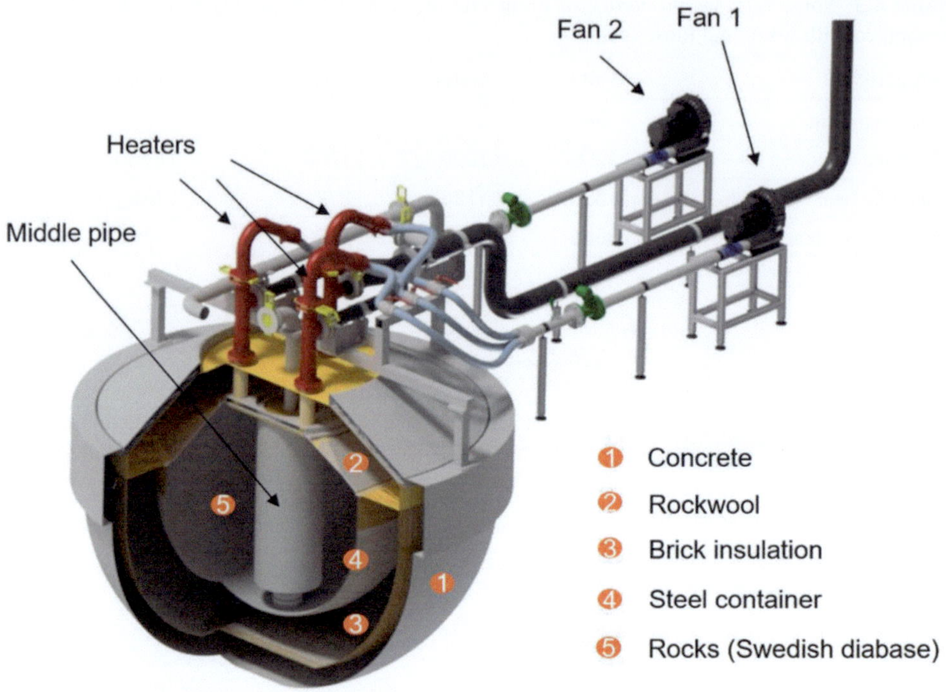

Fig. 4.24 Design of a rock bed thermal energy storage unit. Figure 2 Reprinted from Muhammad et al. (2023) Copyright (2023) with permission from Elsevier. CC BY 4.0. https://creativecommons.org/licenses/by/4.0/

References

Ballzus C, Frimmannson H, Gunnarsson G I, Hrolfsson I (2000) The geothermal power plant at Nesjavellir, Iceland. In: Proceedings of world geothermal congress, 28 May 2000–10 Jun 2000 Kyushu-Tohoku, Japan

Boesten S, Ivens W, Dekker SC, Eijdems H (2019) 5th generation district heating and cooling systems as a solution for renewable urban thermal energy supply. Adv Geosci 49:129–136. https://doi.org/10.5194/adgeo-49-129-2019

Bowlingbal (2009) Pipeline from Nesjavellir Geothermal Power Plant to Reykjavik. https://web.archive.org/web/20161017164132/http://www.panoramio.com/photo/43782438

Choudhury C, Chauhan PM, Garg HP (1995) Economic design of a rock bed storage device for storing solar thermal energy. Sol Energy 55:29–37. https://doi.org/10.1016/0038-092x(95)00023-k

Csp.guru (2022) China energy engineering corporation project has 8 hours of storage and can deliver power well into the night. https://commons.wikimedia.org/wiki/File:50_MW_molten-salt_power_tower_in_hami.jpg

DKFjernvarmen (2011) Solvarme. https://commons.wikimedia.org/wiki/File:Solpaneler.jpg

ENERGY.GOV (2012) Workers construct one of a dozen molten salt tanks at Solana Plant. https://commons.wikimedia.org/wiki/File:Abengoa_Solar_(7336087392).jpg
Forsberg C, Brick S, Haratyk G (2018) Coupling heat storage to nuclear reactors for variable electricity output with baseload reactor operation. Electr J 31:23–31. https://doi.org/10.1016/j.tej.2018.03.008
Gipe P (2011) Reykjavík geothermal district heating. https://commons.wikimedia.org/wiki/File:Reykjavikur-Iceland-20110523-10.jpg
Krapf H (2008) Nesjavallavirkjun. https://commons.wikimedia.org/wiki/File:2008-05-25_20_Nesjavallavirkjun.jpg
Lund H, Werner S, Wiltshire R, Svendsen S, Thorsen J E, Hvelplund F, Mathiesen BV (2014) 4th generation district heating (4GDH): integrating smart thermal grids into future sustainable energy systems. Energy 68:1–11. https://doi.org/10.1016/j.energy.2014.02.089
Mrmw and Andol (2018) Picture showing the 4 different generations of district heating systems. https://commons.wikimedia.org/wiki/File:Generations_of_district_heating_systems_EN.svg
Muhammad Y, Saini P, Knobloch K, Frandsen HL, Engelbrecht K (2023) Rock bed thermal energy storage coupled with solar thermal collectors in an industrial application: simulation, experimental and parametric analysis. J Energy Storage 67:107349. https://doi.org/10.1016/j.est.2023.107349
Nygaard S (2009) "Under-floor heating" in the street. https://www.flickr.com/photos/stignygaard/4143658508/
PommeDe Therme (2024) Production d'eau chaude sanitaire ECS. https://commons.wikimedia.org/wiki/File:Chaufferie.png
Riedy C, Milne G, Ryan P, Alviano P, Dwyer S (2020) Hot water systems. http://www.yourhome.gov.au/energy/hot-water-service
Rubio-Maya C, Martínez E, Ambriz-Díaz V (2015) Cascade utilization of low and medium enthalpy geothermal resources—a review. Renew Sustain Energy Rev 52:689–716. https://doi.org/10.1016/j.rser.2015.07.162
Sarbu I, Mirza M, Muntean D (2022) Integration of renewable energy sources into low-temperature district heating systems: a review. Energies 15:6523. https://doi.org/10.3390/en15186523
Sibbitt B, McClenahan D, Djebbar R, Thornton J, Wong B Carriere J, Kokko J (2012). The performance of a high solar fraction seasonal storage district heating system—five years of operation. Energy Proc 30:856–865. https://doi.org/10.1016/j.egypro.2012.11.097
Singh H, Saini RP, Saini JS (2013) Performance of a packed bed solar energy storage system having large sized elements with low void fraction. Sol Energy 87:22–34. https://doi.org/10.1016/j.solener.2012.10.004
Strand S (2024) Sun goes down on Drake landing solar community. Okotoks Online (28 June 2024). https://www.okotoksonline.com/articles/sun-goes-down-on-drake-landing-solar-community
Surek S (2006) Solar panels for water heating on top of a hotel in Perissa, Santorini, Greece. https://commons.wikimedia.org/wiki/File:Solar_hot_water_panels,_Santorini.jpg
U. S. Department of Energy (2016) Linear concentrator solar power plant illustration. https://www.energy.gov/eere/solar/articles/linear-concentrator-solar-power-plant-illustration

5 Solar Ponds

5.1 Introduction

Since solar energy is not constant in time, its effective use requires suitable energy storage facilities. Several of the most important of these have been discussed in previous chapters. These include pumped hydroelectric and compressed air, both of which are effective methods of storing electrical energy from photovoltaics or thermal solar generation. Sensible heat storage is a suitable method of storing solar thermal energy directly using the heat capacity of water, soil, rock or molten salt. Solar ponds are unique approach to storing and utilizing solar thermal energy. A solar pond not only serve as a solar thermal storage device but also act as the solar collector itself. While the overall efficiency of a solar pond is quite low, as discussed below, its large size and simple technology allows it to be a cost-effective means of utilizing solar energy for a number of applications. There are several solar pond designs that have been developed. The present chapter concentrates on the most common of these, the salinity gradient solar pond.

5.2 Basic Principles of Solar Ponds

A solar pond is an artificial pond that is analogous to a flat plate solar collector coupled to a thermal water storage tank. That is, it converts solar radiation into heat and stores the resulting thermal energy. The development of solar ponds follows from the observation that some natural salt lakes develop large thermal and salinity gradients during the summer months. In some cases, surface water may be nearly fresh at a temperature of around 25 °C, while water at a depth of 1–2 m may be a nearly saturated salt solution at a temperature of up to 70 °C. This phenomenon may be applied to the construction of an

R. A. Dunlap, *Renewable Energy Storage*, Synthesis Lectures on Renewable Energy Technologies, https://doi.org/10.1007/978-3-031-88942-4_5

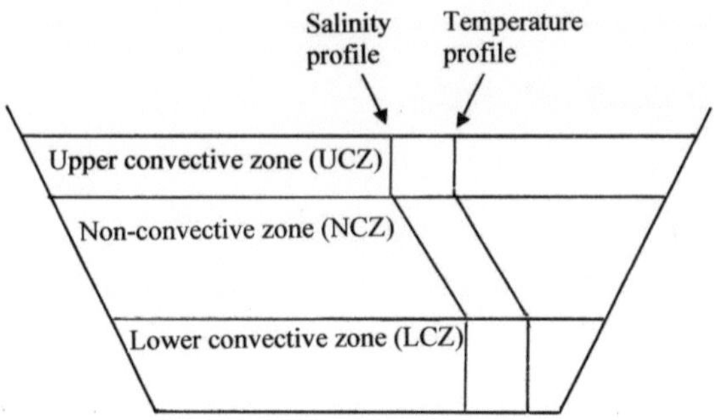

Fig. 5.1 Diagram of the basic features of a solar pond

artificial solar pond where energy is acquired from solar radiation and is stored as a result of the natural convective structure of the pond.

Figure 5.1 shows the relevant features of a simple salinity gradient solar pond. The pond is divided vertically into three distinct zones. The zone at the top of the pond is the upper convective zone (UCZ) where the temperature and salinity are approximately constant. The middle zone is the non-convective zone (NCZ) where there are both temperature and salinity gradients. Both temperature and salinity increase with increasing depth below the surface in the non-convective zone. The bottom zone is the lower convective zone (LCZ) where, again, the temperature and salinity are more or less constant. Typically, the lower convective zone is a concentrated brine solution and heat is stored (and extracted) from this zone. The non-convective zone acts to insulate the warmer lower convective zone from the cooler upper convective zone. In the construction of a solar pond, it is important to utilize an impermeable (e.g., plastic) liner to avoid salt loss into the soil and the associated environmental contamination.

Solar radiation which is incident upon the pond interacts first with the upper convective zone. Some of the radiation will be reflected from the surface and will not contribute to the heat content of the pond. Some of the incident solar radiation will be absorbed as heat in the upper convective zone. Most of this is lost back to the atmosphere through radiation or convection. A fraction of the incident solar radiation will be transmitted through the upper layer and will be absorbed in the non-convective zone. Some will also pass through the non-convective zone and be absorbed as heat in the lower convective zone. Heat which is absorbed in the lower convective zone is trapped because of the insulating non-convective zone above it. The resulting temperature difference between the upper and lower convective zones can be as large as 60 °C. It is the sensible heat that is trapped in the lower convective zone that is stored and that can be recovered.

As the lower convective zone is insulated from the cooler zones above it and serves as the medium for sensible heat storage, it is possible add additional heat to this zone for storage. This can be done, as illustrated in Fig. 5.2, by pumping hot water from external solar collectors, in this case evacuated tube collectors, into this zone.

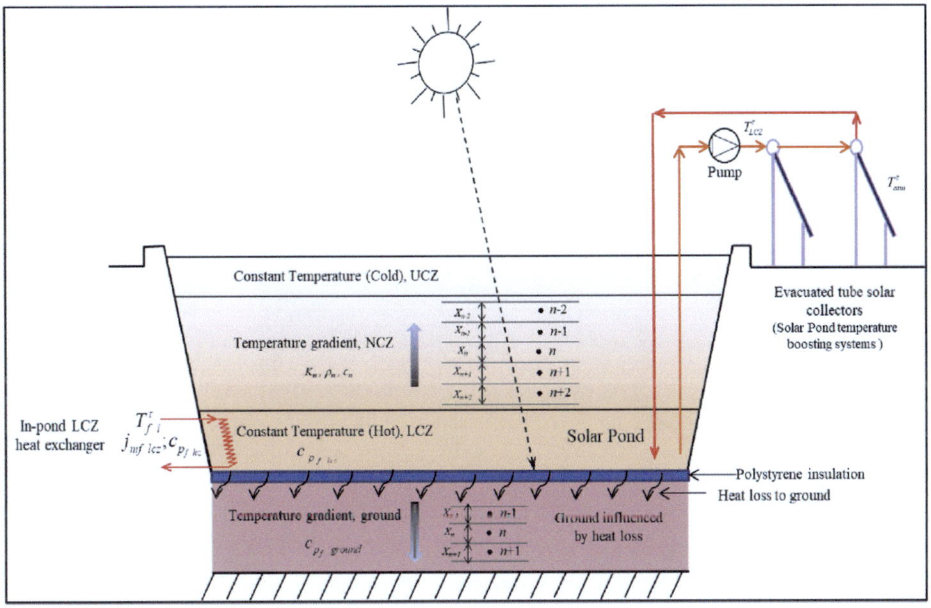

Fig. 5.2 Solar pond showing the addition of heat in the lower convective zone from external evacuated tube solar collectors. Figure 2 Reprinted from Ganguly et al. (2019) Copyright (2018) by The Authors, Published by Elsevier Ltd. CC BY 4.0. https://creativecommons.org/licenses/by/4.0/

5.3 Applications of Solar Ponds

Some of the applications of thermal energy that is stored in a solar pond are summarized in Fig. 5.3. These applications involve the direct use of the thermal energy content of the pond and the production of electrical energy through the use of a heat engine. In the present section, we provide a brief introduction to some of the most common and important applications of solar ponds. We begin with a general discussion of how stored heat can be removed from a solar pond.

5.3.1 Solar Pond Heat Removal

The simplest method of utilizing heat from a solar pond is to use an external heat exchanger as illustrated in Fig. 5.4a. Hot concentrated brine is pumped from the lower convective layer through an external heat exchanger where the heat content is transferred to water flowing through the heat exchanger. After passing through the heat exchanger, the brine is returned to the lower convective zone. The heat extracted by the heat exchanger can be used for various applications as described in Fig. 5.3. The potential difficulty of

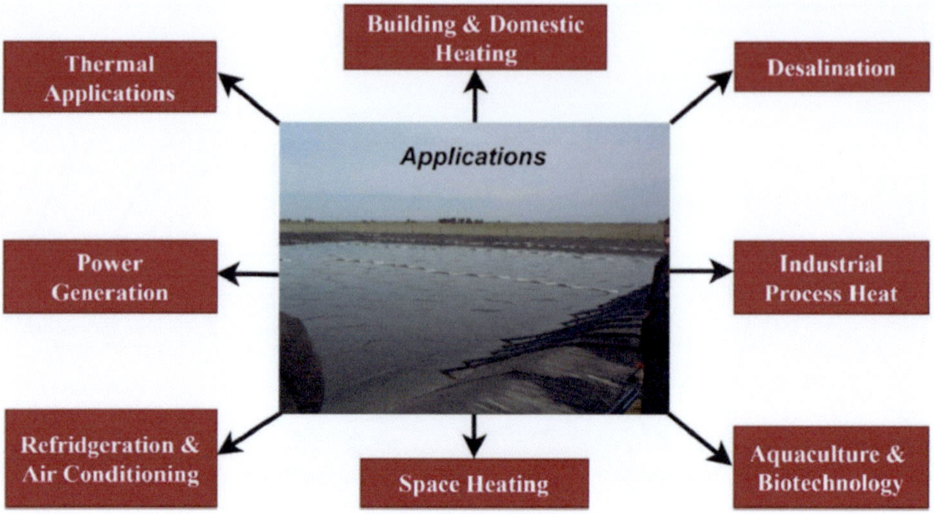

Fig. 5.3 Typical applications of heat stored in a solar pond. Figure 9 from Mbelu et al. (2024) With permission of Springer. CC BY 4.0. https://creativecommons.org/licenses/by/4.0/

this approach is that the brine which is returned to the lower convection zone after passing through the external heat exchanger is colder, and therefore denser than the hot brine already in this zone and tends to lie on the bottom without mixing efficiently with the warmer brine in this zone.

Another method of extracting heat from the solar pond is through the use of an internal heat exchanger as illustrated in Fig. 5.4b. Here cold fresh water is circulated through a heat exchanger located in the lower convective zone. The fresh water is heated and can then be used (e.g.) for space heating, either as hot water or the heat can be transferred to air with an external water-to-air heat exchanger, as shown in the figure. While this alleviates the problem related to mixing associated with the external heat exchanger, the system is more complex, and the location of the internal heat exchanger makes maintenance more involved.

Recent studies (Leblanc et al. 2011) have shown that the overall thermal efficiency may be improved by also utilizing the heat content in the non-convective zone. Figure 5.4c shows an arrangement for an internal heat exchanger that utilizes the heat content of the non-convective zone as well as the lower convective zone.

5.3.2 Industrial Heat

Industrial processes use heat for a wide variety of applications. In the past, this heat has typically been provided by the combustion of fossil fuels or electrical resistive heaters.

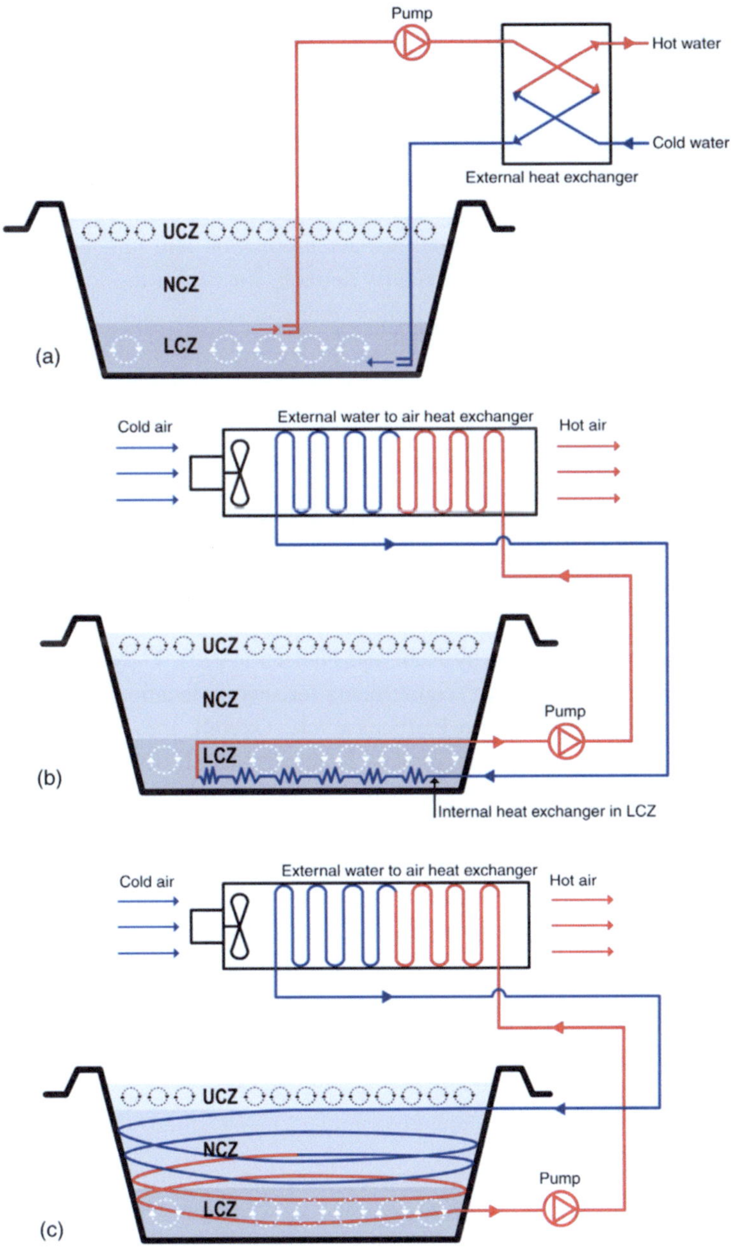

Fig. 5.4 Method of extracting heat from a solar pond using **a** an external heat exchanger, **b** an internal heat exchanger and **c** an internal heat exchanger in the non-convective gradient zone and the lower convective zone. Figure 14.3 Reprinted from Valderrama et al. (2016) Copyright (2016) with permission from Elsevier

Some of the required heat is in the 100–200 °C range, which is beyond the range of temperatures that can be achieved in a solar pond (although there are some possible approaches to dealing with this, as discussed in Sect. 5.3.4). However, a number of industrial processes utilize heat in the range of 80 °C and the use of thermal energy recovered from a solar pond can be a beneficial contribution to greenhouse gas reduction. Following the diagrams in Fig. 5.4, it is seen that the solar pond can produce hot water or hot air for use in industrial processes. Heat from previously constructed solar ponds has been successfully utilized for processes such as salt production and lumber drying. Thermal energy from solar ponds is also particularly suitable for use in the food industry. These applications include drying fruit and grains, dairy processes and canning (Kumar and Kishore 1999).

5.3.3 Space Heating

The use of solar ponds for space heating follows along similar lines to their application for industrial process heat described above. For this purpose, the very large thermal mass associated with the heat storage means that heat can be available day and night and independent of weather conditions or season. The application of a solar pond for residential space heating has been reviewed by Rabl and Nielsen (1975). They have looked at solar pond thermal capacity and heating requirements for several locations in the United States ranging from Albuquerque, NM to Fairbanks, AK. Overall, they conclude that in temperate climates, the solar pond can provide adequate heating when it is comparable in surface area and volume to the building to be heated. For example, for a small house that is approximated by a box 5 m × 10 m × 15 m, a suitable solar pond would be 550 m^2 by 1.4 m deep. It has been suggested that solar ponds can be effectively operated in conjunction with a heat pump. In the winter, the heat pump can complement the solar pond thermal output. In the summer the heat pump can operate as an air conditioner, where the upper convection zone of the solar pond can serve as a cold reservoir to improve the coefficient of performance of the air conditioner (El-Sebaii et al. 2011).

5.3.4 Desalination

A common application for the thermal energy stored in a solar pond is seawater desalination. There are several approaches to seawater desalination. One of the most effective technologies is multi-stage flash desalination as shown by the diagram in Fig. 5.5. The principle of operation of the multi-stage flash desalinator has been described by Saleh et al. (2011) and Warsinger et al. (2015). Cold seawater is pumped into the device (at the upper right in the figure) and is heated as it progresses through the stages of the desalinator. After passing through the various stages of the desalinator, the seawater (at this point

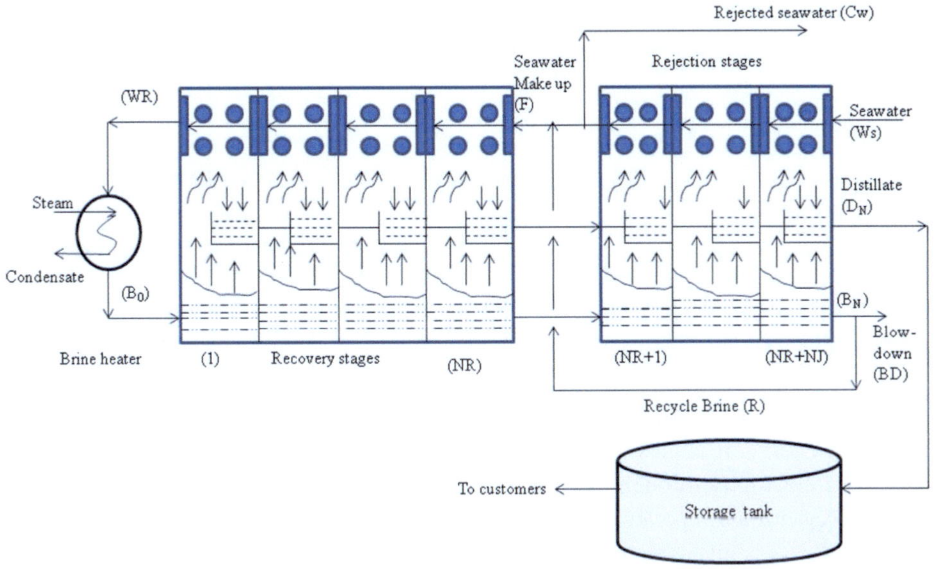

Fig. 5.5 Schematic of a multistage flash evaporator for desalination of seawater. Figure 3 from Said et al. (2013) Copyright 2013 by the Author CC BY 3.0. http://creativecommons.org/licenses/by/3.0/

referred to as brine) passes through the brine heater and the temperature is increased by a small additional amount. After passing through the heater the brine is passed back through the stages of the desalinator where the temperature and pressure progressively decrease. At each stage, the temperature of the brine is slightly above its boiling point and a portion of the brine is flash boiled and this reduces the temperature of the remaining brine before it passes to the next stage. The steam condenses and contributes to the flow of desalinated water through the device.

As the temperature of the water extracted from a solar pond is typically in the range of about 80 °C, it may not be ideal for providing heat to the brine heater of the multi-stage flash desalinator. It has been suggested that an absorption heat transformer could be used as a means of improving the performance of the desalinator (Salata and Coppi 2014). An absorption heat transformer transfers heat from an intermediate temperature source to a high temperature reservoir and to a low temperature reservoir. That is, it concentrates the heat flow to one stream and depletes it from another. The basic principle of the absorptive heat transformer is shown in Fig. 5.6. This approach would allow desalinators, or other applications, that require higher temperature thermal input to utilize the intermediate temperature output from a solar pond.

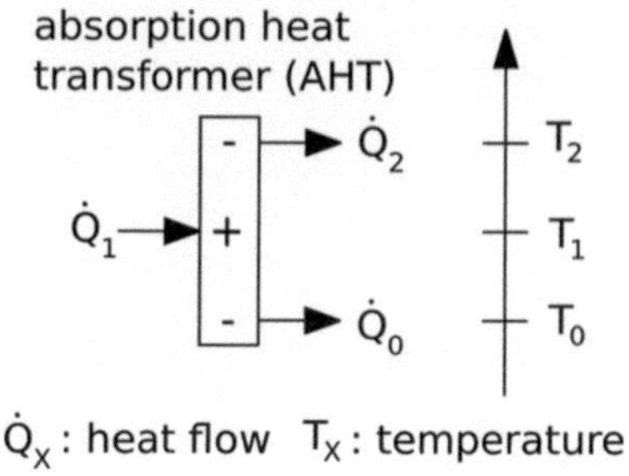

Fig. 5.6 Basic principle of an absorptive heat transformer. Fcudok (2018) CC BY-SA 4.0. https://creativecommons.org/licenses/by-sa/4.0/deed.en

5.3.5 Electricity Generation

The applications of solar pond energy as described above deal with the direct utilization of thermal energy. It is also possible to use the thermal energy output of the solar pond to produce electricity through the use of a heat engine. The ideal Carnot efficiency of a heat engine with a temperature of the hot reservoir, T_{hot}, and a temperature of the cold reservoir, T_{cold}, is given by,

$$\eta = 100 \times \frac{T_{hot} - T_{cold}}{T_{hot}}. \tag{5.1}$$

Using $T_{hot} = 80\ °C = 353$ K and $T_{cold} = 20\ °C = 293$ K, gives an ideal Carnot efficiency of

$$\eta = 100 \times \frac{353\ \text{K} - 293\ \text{K}}{353\ \text{K}} = 17\%. \tag{5.2}$$

In reality actual efficiencies of solar pond electricity generation is more in the range of about 3%. It is sometimes useful to define the solar pond energy yield as the ratio of the pond thermal or electrical energy production and the collected solar energy.

Figure 5.7 shows the design of a typical system for generating electricity from the thermal output of a solar pond. Obviously, temperatures above 100 °C are necessary to produce steam, and typical coal-fired thermal generating stations utilize steam at around 540 °C. The thermal output of the solar pond is at too low a temperature for the utilization of a steam turbine. In view of the available temperatures, it is appropriate to use an organic Rankine cycle turbine. The organic Rankine cycle uses an organic fluid with a boiling point that is lower than that of water. This allows the heat from the solar pond to be used directly to produce vapor from the organic fluid through the evaporator, as shown in the figure. Typical organic fluids that are suitable for this application include freon and propane. Arjunan et al. (2022) have recently provided an extensive review of the applicability of various organic fluids for Rankine cycle heat engines.

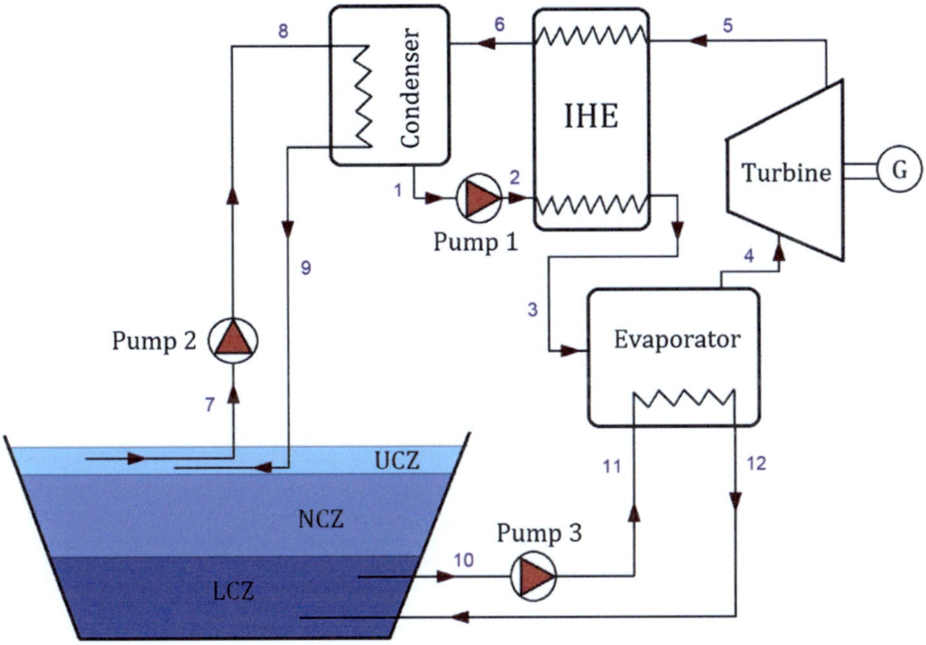

Fig. 5.7 Method of using heat from a solar pond for the generation of electricity by means of an organic Rankine cycle turbine. *IHE* internal heat exchanger. Figure 2 Reprinted from Ziapour and Shokrnia (2017) Copyright (2017) with permission from Elsevier

5.4 History of Solar Ponds

A naturally occurring form of solar pond was discovered in the 1940s at Medve Lake in Transylvania (Hungary), where temperatures of up to 70 °C were observed at a depth of 1.32 m at the end of summer (El-Sebaii et al. 2011). Other naturally occurring salt lakes which provide thermal energy storage in a lower convective zone have been discovered such as those in the Atacama Desert Plateau of South America, situated between the Pacific Ocean and the coastal range of mountains. Interest in the construction of artificial solar ponds began in the mid-twentieth century (Tabor and Matz 1965). Some of the more significant solar ponds that have been constructed over the years are described below.

5.4.1 Beit HaArava Solar Pond, West Bank

One of the earliest solar ponds and the largest constructed to date, was built in Beit HaArava, an Israeli settlement and kibbutz in the West Bank in 1983. This 210,000 m^2 pond produced 5 MW electrical output and was the first solar pond to be successfully used for electricity generation. The organic Rankine turbines were developed by Ormat Industries for the purpose of utilizing intermediate temperature solar pond thermal resources for electricity generation. While the operation of the Beit HaArava solar pond demonstrated the technical feasibility of electricity generation from solar pond thermal output, it was not commercially viable and was shut down in 1988.

5.4.2 El Paso Solar Pond, Texas

The El Paso solar pond was developed as a research and demonstration project of the University of Texas at El Paso (Lu and Swift 2001). It had a surface area of about 3000 m^2 and a depth of 3.2 m. The project began in 1983, and the pond was constructed on the property of Bruce Foods, Inc., a canning company specializing in Mexican foods. The El Paso solar pond was the first salinity gradient solar pond in the United States and, in 1985, became the first in the world to provide process heat to an industrial user (Bruce Foods). In 1986, it became the first solar pond in the United States to produce electricity. An organic Rankine turbine generated an electrical output of 100 kW that was fed into the local grid and provided power for Bruce Foods. In 1987, the El Paso solar pond became the first solar pond in the United States to be operated in conjunction with a desalination facility (Lu et al. 2001). The approach was to produce desalinated water and to reuse the concentrated brine solution to construct additional solar ponds. The pond operated until 2003.

5.4.3 Bhuj Solar Pond, Gujarat, India

The Bhuj solar pond was constructed at the Kutch Dairy in Bhuj, India as a collaborative project between the Gujarat Dairy Development Corporation, the Gujarat Energy Development Agency and the Tata Energy Research Institute. Construction began in 1987, and the pond became operational in 1993. To reduce construction cost, the pond utilized a liner that consisted of alternating layers of clay and LDPE (low density polyethylene). It was the first solar pond in India to supply process heat to an industrial user. The 6000 m^2 solar pond provided heat to the dairy for milk processing from 1993 to 1995 and, after about a year break, from 1996 to 2000. In 2000, the Gujarat Dairy Development Corporation was adversely affected by economic losses and, shortly thereafter, the Kutch Dairy suffered severe earthquake damage. The operation of the Bhuj solar pond was subsequently discontinued.

Fig. 5.8 The 3000 m^2 solar pond at Pyramid Hill, Australia. Figure 3 Reprinted from Malik et al. (2011) Copyright (2011) with permission from Elsevier

5.4.4 Pyramid Hill, Victoria, Australia

A 3000 m^2 solar pond, constructed in 2001 in Pyramid Hill, Australia, provides thermal energy that is used for commercial salt production by Pyramid Salt Pty Ltd. (Bawahab et al. 2019). A heat pump, powered by a solar photovoltaic array, is used in conjunction with the heat output of the solar pond to preheat inlet air from ~ 20 to ~ 70 °C before being electrically heated to 80–100 °C for salt drying. The incorporation of the thermal heat from the solar pond reduces electricity consumption for air heating by 50%. A photograph of the Pyramid Hill solar pond is shown in Fig. 5.8. The photograph shows that the surface of the solar pond is covered with floating plastic rings, each about 2 m in diameter. The rings are designed to minimize detrimental effects from wind driven waves (see analysis by Akbarzadeh et al. 1983).

5.4.5 Granada Solar Pond, Spain

The Granada solar pond was constructed in 2014 at the Solvay Minerales facility in Granada, Spain (Alcaraz et al. 2018a, b; Montalà et al. 2022) as a joint project between Solvay Iberica, the Polytechnic University of Catalonia and the Royal Melbourne Institute of Technology (RMIT University). The solar pond has an area of 500 m^2 and a depth of 2.2 m and is located at a mining facility that produces $SrSO_4$ (celestine). Celestine is used as a source of strontium and is important in the manufacture of ceramic magnets and flares, glassmaking and medical treatment of conditions such as tooth sensitivity, osteoporosis, osteoarthritis and prostate cancer. The processed ore has a celestine content of 30–50%. This is increased to around 90% by milling followed by froth flotation. Froth flotation is a common process in mineral processing where valuable minerals are separated from gangue (the worthless material that is mixed with the mineral in the ore) by differences in their hydrophobicity, that is the degree to which the material is attracted to or repelled by water. The flotation process requires temperatures ~ 60 °C, and thermal energy from the solar pond is used to supplement heat from a fuel oil-fired boiler to achieve the necessary temperature. It has been estimated that the incorporation of the thermal energy from the solar pond reduces the fuel consumption by more than 50%.

References

Akbarzadeh A, MacDonald R, Wang Y (1983) Reduction of surface mixing in solar ponds by floating rings. Sol Energy 31:377–380. https://doi.org/10.1016/0038-092X(83)90136-6

Alcaraz A, Montalà M, Cortina JL, Akbarzadeh A, Aladjem C, Farran A, Valderrama C (2018a) Design, construction, and operation of the first industrial salinity-gradient solar pond in Europe: an efficiency analysis perspective. Sol Energy 164:316–326. https://doi.org/10.1016/j.solener.2018.02.053

Alcaraz A, Montalà M, Valderrama C, Cortina J, Akbarzadeh A, Farran A (2018b) Thermal performance of 500 m^2 salinity gradient solar pond in Granada, Spain under strong weather conditions. Sol Energy 171:223–228. https://doi.org/10.1016/j.solener.2018.06.072

Arjunan P, Gnana Muthu JH, Somanasari Radha SL, Suryan A (2022) Selection of working fluids for solar organic Rankine cycle—a review. Int J Energy Res 46:20573–20599. https://doi.org/10.1002/er.7723

Bawahab M, Faqeha H, Ve QL, Faghih A, Date A, Akbarzadah A (2019) Industrial heating application of a salinity gradient solar pond for salt production. Energy Proc 160:231–238. https://doi.org/10.1016/j.egypro.2019.02.141

El-Sebaii A, Ramadan M, Aboul-Enein S, Khallaf A (2011) History of the solar ponds: a review study. Renew Sustain Energy Rev 15:3319–3325. https://doi.org/10.1016/j.rser.2011.04.008

Fcudok (2018) The driving heat flow Q_1 at intermediate temperature level T_1 will be split in the revalued heat flow Q_2 at high temperature level T_2 and in rejected heat flow Q_0 at low temperature level T_0. https://commons.wikimedia.org/wiki/File:Blackbox_AHT.png

Ganguly S, Date A, Akbarzadeh A (2019) On increasing the thermal mass of a salinity gradient solar pond with external heat addition: a transient study. Energy 168:43–56. https://doi.org/10.1016/j.energy.2018.11.090

Kumar A, Kishore VVN (1999) Construction and operational experience of a 6000 m^2 solar pond at Kutch, India. Sol Energy 65:237–249. https://doi.org/10.1016/s0038-092x(98)00134-0

Leblanc J, Akbarzadeh A, Andrews J, Lu H, Golding P (2011) Heat extraction methods from salinity-gradient solar ponds and introduction of a novel system of heat extraction for improved efficiency. Sol Energy 85:3103–3142. https://doi.org/10.1016/j.solener.2010.06.005

Lu H, Swift AHP (2001) El Paso solar pond. J Sol Energy Eng 123:178. https://doi.org/10.1115/1.1384572

Lu H, Walton JC, Swift AHP (2001) Desalination coupled with salinity-gradient solar ponds. Desalination 136:13–23. https://doi.org/10.1016/S0011-9164(01)00160-6

Malik N, Date A, Leblanc J, Akbarzadeh A, Meehan B (2011) Monitoring and maintaining the water clarity of salinity gradient solar ponds. Sol Energy 85:2987–2996. https://doi.org/10.1016/j.solener.2011.08.040

Mbelu OV, Adeyinka AM, Yahya DI, Adediji YB, Njoku H (2024) Advances in solar pond technology and prospects of efficiency improvement methods. Sustain Energy Res 11:18. https://doi.org/10.1186/s40807-024-00111-5

Montalà M, Ganesan K, Casal O, Cortina J, Santarelli M, Valderrama C (2022) Energy, exergy and thermoeconomic analysis of an industrial solar pond. Sol Energy 242:143–156. https://doi.org/10.1016/j.solener.2022.07.014

Rabl A, Nielsen CE (1975) Solar ponds for space heating. Sol Energy 17:1–12. https://doi.org/10.1016/0038-092X(75)90011-0

Said SA, Emtir M, Mujtaba IM (2013) Flexible design and operation of multi-stage flash (MSF) desalination process subject to variable fouling and variable freshwater demand. Processes 1:279–295. https://doi.org/10.3390/pr1030279

Salata F, Coppi M (2014) A first approach study on the desalination of sea water using heat transformers powered by solar ponds. Appl Energy 136:611–618. https://doi.org/10.1016/j.apenergy.2014.09.079

Saleh A, Qudeiri JA, Al-Nimr MA (2011) Performance investigation of a salt gradient solar pond coupled with desalination facility near the Dead Sea. Energy 36:922–931. https://doi.org/10.1016/j.energy.2010.12.018

Tabor H, Matz R (1965) A status report on a solar pond project. Sol Energy 9:177–182. https://doi.org/10.1016/0038-092X(65)90044-7

Valderrama C, Cortina JL, Akbarzadeh A (2016) Solar ponds, chap 14. In: Letcher TM (ed) Storing energy—with special reference to renewable energy sources. Elsevier, Amsterdam, pp 273–289. https://doi.org/10.1016/B978-0-12-803440-8.00014-2

Warsinger DM, Mistry KH, Nayar KG, Chung HW, Lienhard JHV (2015) Entropy generation of desalination powered by variable temperature waste heat entropy. Entropy 17:7530–7566. https://doi.org/10.3390/e17117530

Ziapour BM, Shokrnia M (2017) Exergoeconomic analysis of the salinity-gradient solar pond power plants using two-phase closed thermosyphon: a comparative study. Appl Therm Eng 115:123–133. https://doi.org/10.1016/j.applthermaleng.2016.12.129

6 Latent Heat and Thermochemical Energy Storage

6.1 Introduction

In the previous chapter, we saw how energy could be stored as sensible heat in liquids, such as water, or solids, such as rock. In the present, chapter we look at how the latent heat associated with phase transitions, or the thermochemical energy associated with sorption or chemical reactions can be used to store thermal energy. Figure 6.1 shows a breakdown of the categories of materials for the storage of thermal energy. Phase change materials (PCM) can store thermal energy associated with the latent heat of fusion for a solid–liquid transition or the latent heat of vaporization for liquid–gas transition. Thermochemical energy associated with sorption, that is absorption or adsorption processes, can provide a means of energy storage. The heat associated with reversible chemical reactions can also be utilized for energy storage. The basic processes associated with these phenomena along with some examples of their applications are described in the present chapter.

6.2 Basics of Latent Heat Energy Storage

Latent heat energy storage takes advantage of the large amount of heat that accompanies phase changes in a material. Typical examples of phase transitions are the transitions between the solid, liquid and gaseous forms of a material. The energy associated with the phase transition between the solid state and the liquid state (melting) is the latent heat of fusion and the energy associated with the phase transition between the liquid state and the gaseous state is the latent heat of vaporization. Figure 6.2 shows the temperature of a material a function of the energy input and illustrates the latent heat associated with the phase transitions.

R. A. Dunlap, *Renewable Energy Storage*, Synthesis Lectures on Renewable Energy Technologies, https://doi.org/10.1007/978-3-031-88942-4_6

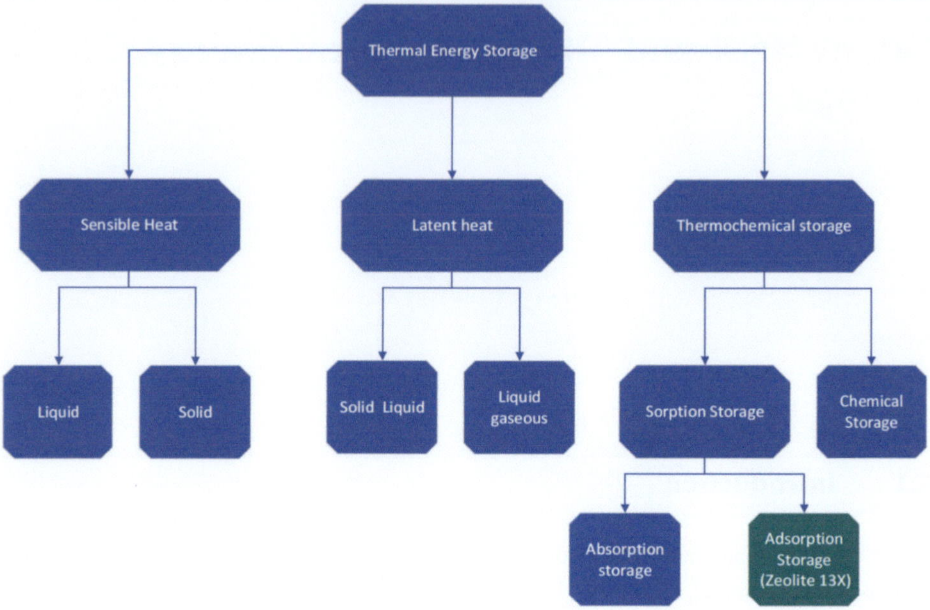

Fig. 6.1 Categories of thermal energy storage. Figure 1 from Banaei and Zanj (2021) Copyright 2021 by the Author CC BY 4.0. https://creativecommons.org/licenses/by/4.0/

Fig. 6.2 Temperature as a function of energy input for a material showing the latent heat of fusion and the latent heat of vaporization

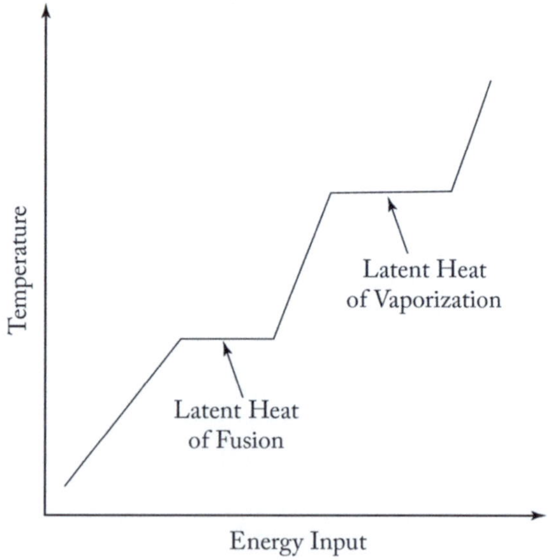

The large change in the heat content of a material associated with a phase transition at a temperature, T, may be explained in terms of the Gibbs free energy, G, which is defined as

$$G = H - TS. \tag{6.1}$$

Here H is the heat content of the material and S is the entropy. At a phase transition, the change in the Gibbs free energy is

$$\Delta G = \Delta H - T\Delta S, \tag{6.2}$$

where, as shown in Fig. 6.1, the temperature remains constant. As the Gibbs free energy at a phase transition is a continuous function of temperature, the change in Gibbs free energy must be zero. This means that

$$\Delta H = T\Delta S, \tag{6.3}$$

and this shows that, as the entropy (degree of disorder) changes when a material undergoes a change from solid to liquid or from liquid to gas, there is a change in its heat content.

The thermal behavior described above, can provide an efficient means of storing thermal energy. Figure 6.3 compares the sensible heat of a material with the latent heat of (e.g.) fusion of a material that undergoes a solid–liquid phase transition. To store and release significant thermal energy from the material that does not show any phase transition and is characterized by the *sensible heat* line in the figure would require the cycling of the material over a wide range of temperatures. The phase change material, on the other hand, can store and release a greater amount of thermal energy by cycling over a small temperature range from a few degrees below the phase transition to a few degrees above the transition.

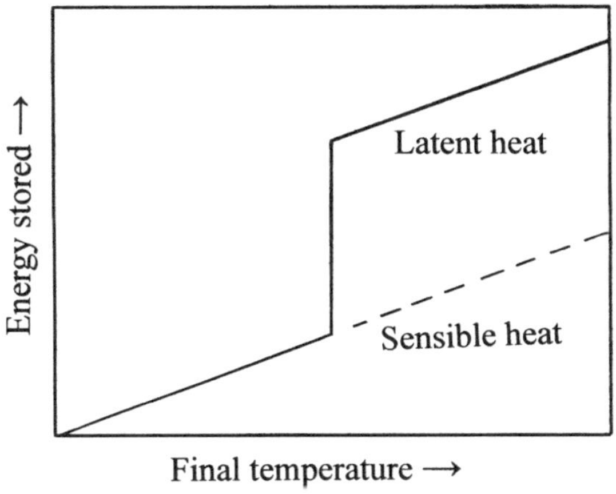

Fig. 6.3 Diagram showing energy stored as a function of final temperature for a material that does not show a phase transition and stores energy as sensible heat and a phase change material (PCM) that stores both sensible heat and the contribution from the latent heat associated with the phase transition

The use of materials which undergo a phase transition (e.g. melting) within the temperature range of the operation of the storage facility have certain advantages over the use of sensible heat storage as discussed in Chap. 4. By properly matching the thermal properties of the storage material with the operating temperature range of the storage system and the heating requirements, some potential advantages may be seen. These include

- use of a smaller mass and/or volume of thermal storage material and
- use of a smaller range of operating temperatures.

Two common examples of the use of latent heat for thermal energy storage, melting and vaporization are discussed below.

6.3 Solid–Liquid Systems

6.3.1 Latent Heat Energy Storage in Ice

Table 6.1 shows a comparison of the latent heats associated with the phase transitions in water and a comparison with the sensible heat associated with raising the temperature of water from 0 °C to room temperature (20 °C) and from room temperature to 100 °C. It is clear from the table that the latent heat associated with phase transitions can be much larger than the sensible heat associated with temperature changes of the material. This is particularly the case for the latent heat associated with the liquid to gas phase transition.

The use of the latent heat of fusion of water is a convenient means of load leveling in order to provide air conditioning. In warm summer weather, the air conditioning of large buildings represents one of the most significant contributions to electricity use in many cities and load leveling using the latent heat of water can significantly reduce these requirements. Ice is produced by refrigeration techniques in the night when cooling needs are minimal. At this time, electricity demand is minimal and, as well, electricity rates are low. The ice is then used to cool air to provide air conditioning during the day when electricity rates are high and the need for air conditioning is the greatest. This approach has two benefits; it provides a more economic means of air conditioning a building by

Table 6.1 A comparison of the latent heats and the sensible heat of water

Heat	Value (kJ/kg)
Latent heat of fusion	334
Sensible heat 0–20 °C	84
Sensible heat 20–100 °C	335
Latent heat of vaporization	2260

Quantities are in kJ per kg of water

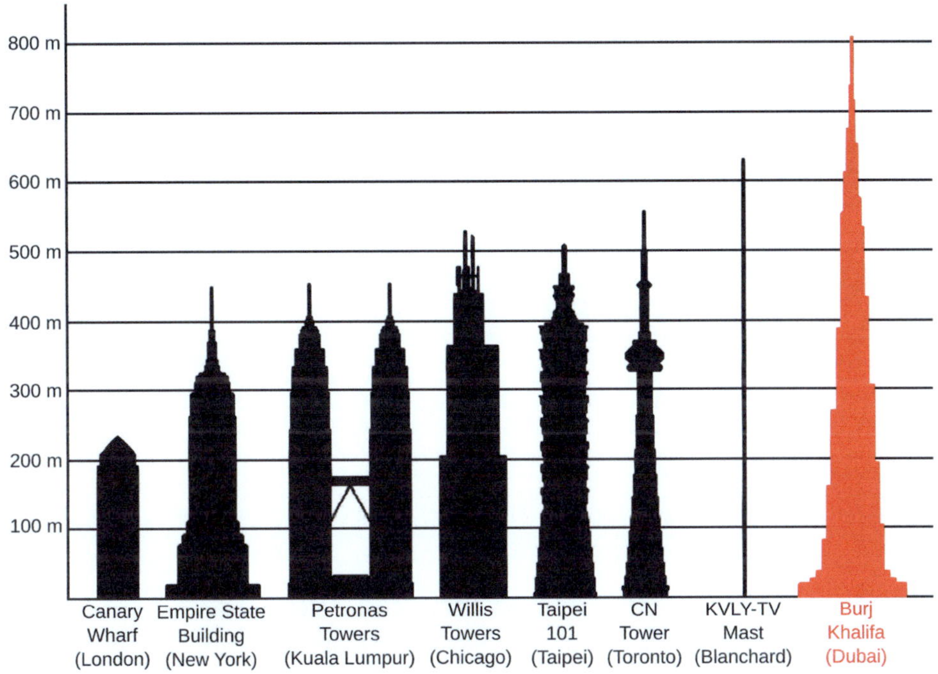

Fig. 6.4 Height of the Burj Khalifa compared to some other well known tall structures. Astronaut (2014) CC BY-SA 3.0. https://creativecommons.org/licenses/by-sa/3.0/deed.en

utilizing electricity at low-rate times and it reduces the need for additional grid generating capacity to cover periods of high demand.

This technology is probably best suited for large office buildings and similar structures. Many major buildings worldwide utilize this technology, including the world's tallest building, Burj Khalifa in Dubai, United Arab Emirates (829.8 m in height), see Fig. 6.4.

6.3.2 Other Solid–Liquid Phase Change Materials

The above example of phase changes in water is only one of the possible approaches to using latent heat for energy storage. The latent heat of fusion associated with the melting transition of other materials may also be useful for the purpose of energy storage and some examples of potential materials are shown in Table 6.2 (see review by Cabeza et al. 2011). The availability of various suitable materials provides the ability to choose a melting temperature and thermal properties that are compatible with a particular application.

The beneficial heat of fusion associated with the solid–liquid transition is shown in Fig. 6.5. Here the thermal storage capacity of octadecane is compared to that of water

Table 6.2 Melting temperatures, densities and heats of fusion for some materials

Material	Melting temperature (°C)	Density (kg/m^3)	Heat of fusion (kJ/kg)
Polyethylene glycol 400	8	1130	100
Caprylic acid	16	910	149
Octadecane	28	777	242
Decanoic acid	32	893	153
Dodecanoic acid	43	930	184
Myristic acid	54	862	196
Paraffin	64	900	174
Stearic acid	69	941	203
NaOH	318	2130	150
$FeCl_2$	670	3160	340
KCl	776	1980	340
NaCl	801	2150	500

over the temperature range from 0 to 50 °C. It is clear that heat stored and recovered (sensible plus latent) from thermal cycling around the transition temperature of 28 °C can provide much improved storage capacity compared to the sensible heat of water over a similar temperature range.

A number of the materials listed in Table 6.2 are organic fatty acids and are suitable for latent heat storage systems because they are non-toxic and are readily available at low cost. Their melting temperature, between room temperature and about 70 °C and fairly high heat of fusion make them suitable for many typical heat storage applications, such as space heating. The mutual solubility of many fatty acids allows for the design of a material with a particular melting temperature. An example of the dodecanoic acid/ myristic acid system is shown in Fig. 6.6, where the melting temperature can be adjusted by the composition from the high value for myristic acid of 54 °C to a low of 34 °C at the eutectic composition of 74 mol% dodecanoic acid.

A notable problem with fatty acids for latent heat energy storage is their typically low thermal conductivity (Naresh et al. 2022). There have been a number of studies of the use of an inorganic matrix to improve the thermal conductivity of fatty acids for thermal energy storage. These inorganic materials include nanostructured samples of carbon, silicon dioxide and copper oxide. The results of a recent study of dodecanoic acid in a fly ash matrix are illustrated in Fig. 6.7. These results show the stability of the resulting composite during repeated thermal cycling.

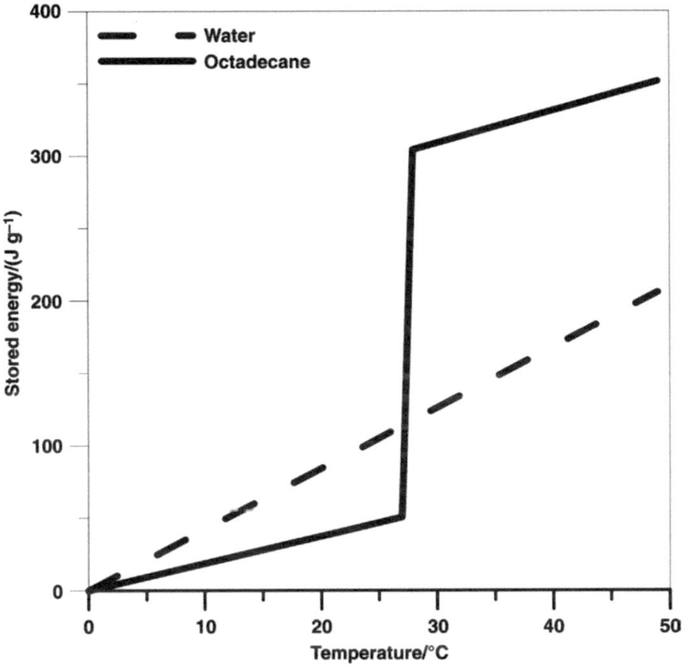

Fig. 6.5 Comparison of sensible heat stored by water and sensible plus latent heat stored by octadecane over the temperature range of 0–50 °C. Figure 22.1 Reprinted from Noël et al. (2022) Copyright (2022) with permission from Elsevier

6.4 Latent Heat Energy Storage in Liquid Air

Another approach to the storage of energy is the use of the latent heat of vaporization of air. In this scheme, excess electricity that is available during the low demand period is used to liquefy air according to the diagram shown in Fig. 6.8 (see Ding et al. 2016). The air is first compressed, which gives rise to the generation of heat as previously described in Chap. 2. This heat is stored for use during the expansion phase of energy recovery. The compressed air is then liquefied and stored.

When additional electricity is required during periods of high demand, the energy stored in the liquid air is recovered as shown in Fig. 6.9. Following along the lines of the energy recovery from compressed air energy storage as described in Chap. 2, the liquid is evaporated and expanded to operate a turbine/generator to produce electricity. During the expansion phase, heat from the compression phase is recovered.

Liquid air energy storage is a potential technology for grid storage to cover periods of peak demand and is, therefore, a possible alternative to pumped hydroelectric storage or traditional compressed air energy storage. In 2011 a small pilot plant providing 350 kW

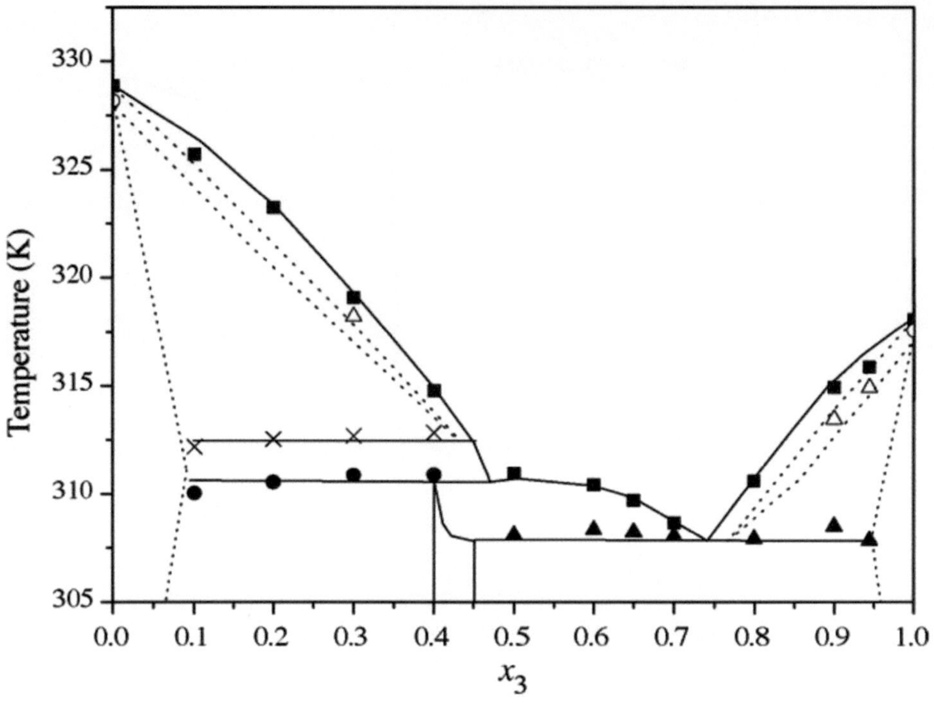

Fig. 6.6 Phase diagram of the dodecanoic acid/myristic acid ($C_{12}H_{24}O_2/C_{14}H_{28}O_2$) system as a function of fraction of dodecanoic acid (x_3). (■) Fusion temperature; (●) peritectic temperature; (▲) eutectic temperature; (×) metatectic temperature; (Δ) transition temperature on the solid phase; (○) transition on the solid phase of the pure component. Figure 12 Reprinted from Costa et al. (2009) Copyright (2009) with permission from Elsevier

output, and a total storage capacity of 2.5 MWh was constructed in Slough, United Kingdom. A larger grid-scale plant providing 5 MW output with a total storage capacity of 15 MWh became operational in Bury, United Kingdom in 2018.

6.5 Thermochemical Energy Storage

As illustrated in Fig. 6.1, thermochemical energy storage can involve reactions or sorption processes, where the sorption processes can be adsorption of adsorption (Kerskes 2016). The distinction between adsorption and absorption is illustrated in Fig. 6.10. Thermochemical energy storage can occur during reactions or sorption processes as illustrated in Fig. 6.11. Energy is stored by dehydration or desorption (charge) by supplying heat to the system. This energy is recovered by hydration or adsorption (discharge) by exposing the system to water molecules.

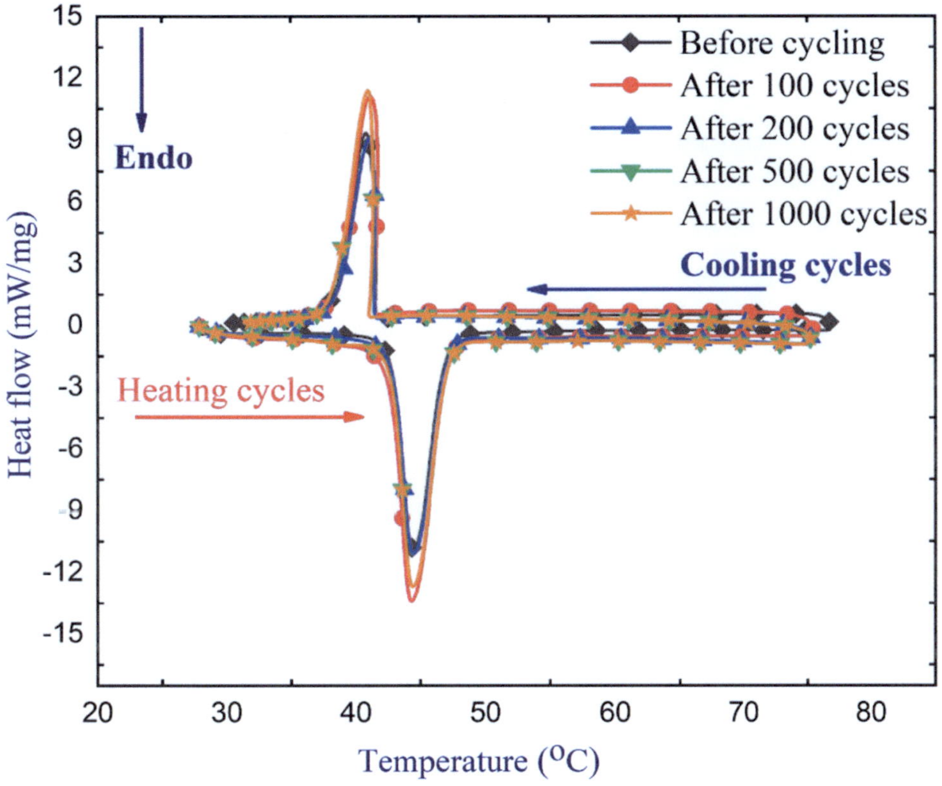

Fig. 6.7 Differential scanning calorimetry scans of the dodecanoic acid/fly ash composite after different numbers of thermal cycles. Figure 9 Reprinted from Naresh et al. (2022) Copyright (2022) with permission from Elsevier. CC BY-NC-ND 4.0. https://creativecommons.org/licenses/by-nc-nd/4.0/

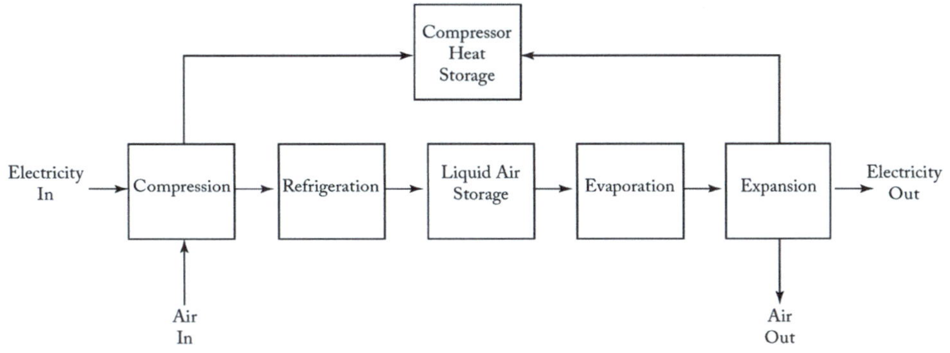

Fig. 6.8 Schematic of liquid air energy storage using the latent heat of vaporization

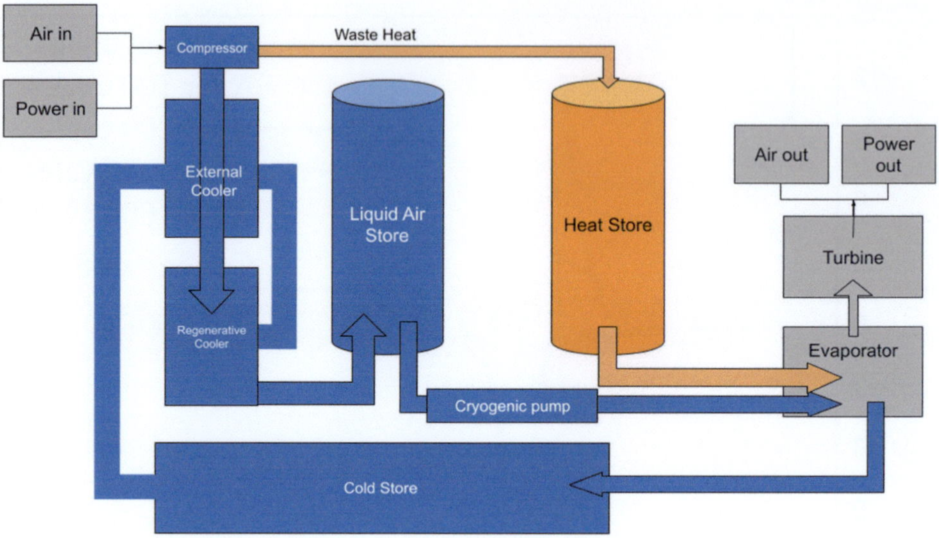

Fig. 6.9 Diagram illustrating flow of air and heat through a cryogenic energy storage system. Dumpygrimbo (2024) CC0 1.0. https://creativecommons.org/publicdomain/zero/1.0/deed.en

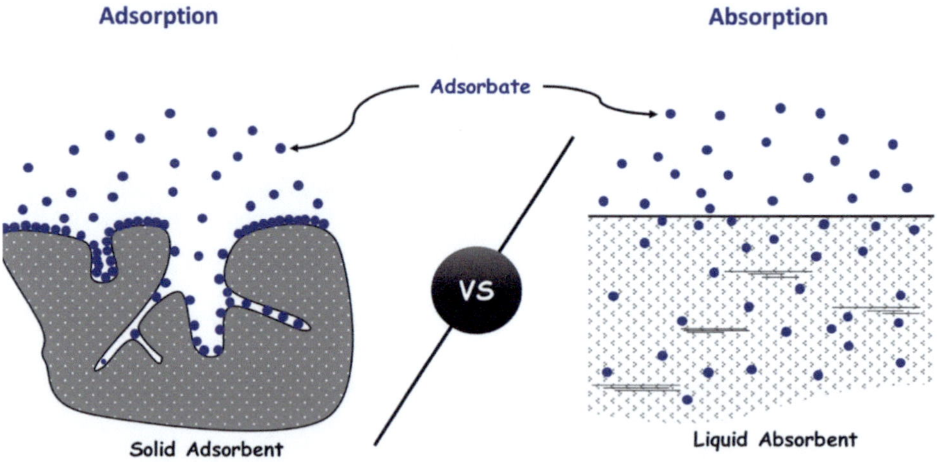

Fig. 6.10 The distinction between adsorption (left) and absorption (right). Figure 3 from Zbair and Bennici (2021) Copyright 2021 by the Authors CC BY 4.0. https://creativecommons.org/licenses/by/4.0/

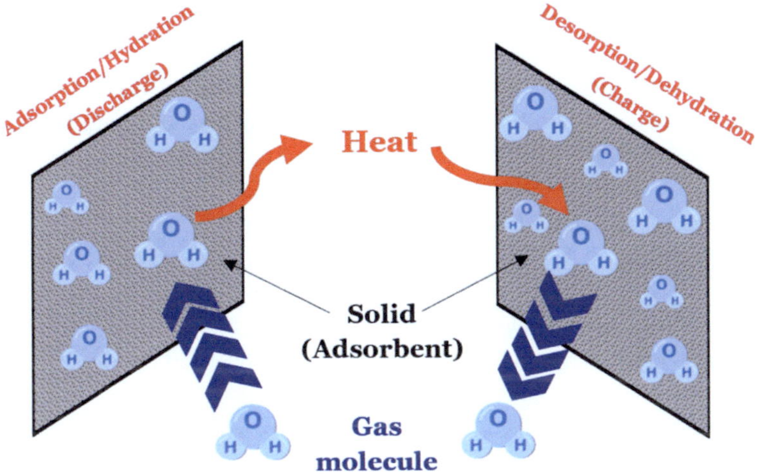

Fig. 6.11 Illustration of energy storage by dehydration or desorption (charge) and energy recovery by hydration or adsorption (discharge). Figure 4 from Zbair and Bennici (2021) Copyright 2021 by the Authors CC BY 4.0. https://creativecommons.org/licenses/by/4.0/

In the two subsections below, we look at examples of solid–liquid reactions and adsorption processes that can be used for thermal energy storage and recovery.

6.5.1 Reactions

Thermal energy can be stored in materials that undergo reversible chemical reactions, as described above. The basic principle follows from the fact that the reaction will be endothermic in one direction and exothermic in the opposite direction. Heat may be stored by causing the endothermic reaction to occur and that heat may be recovered by causing the exothermic reaction to occur. A number of different reactions have been considered for thermochemical energy storage. Although the application of this approach is still in the early stages of development, several types of materials, such as salt hydrates, have shown some promise.

The reversible reaction involving the dehydration and hydration of salt hydrates is of the form

$$\text{salt hydrate} \leftrightarrow \text{salt} + \text{water} \tag{6.4}$$

A typical example of this type of reaction that has been investigated as a possible means of thermochemical energy storage is

$$MgSO_4 \cdot 7H_2O \leftrightarrow MgSO_4 \cdot H_2O + 6H_2O \tag{6.5}$$

The reaction from left to right is endothermic so that thermal energy is stored during the dehydration of the salt hydrate by heating. Energy is recovered during the hydration of the salt by exposing it to water or water vapor, as shown by the exothermic reaction from right to left.

The dehydration of salt hydrates as a means of thermal energy storage has a number of attractive features,

- the materials are common and relatively inexpensive
- the chemical needed for hydration and energy recovery (i.e. water) is universally available
- after dehydration the salt may be stored in a sealed container indefinitely without loss of energy
- theoretical energy storage capacity per unit mass can be as much as 10 times that of sensible heat energy storage in water
- after dehydration the salt may be transported to another location (if appropriate) for hydration (and energy release).

These materials present an attractive alternative to sensible heat storage in water for residential heat storage because of the much smaller mass and volume of material needed. Further research may lead to practical applications for this approach.

6.5.2 Adsorption

A typical reversible adsorption/desorption reaction may be written as (Kerskes 2016; Ristić 2022),

$$A \cdot xH_2O_{(s)} \leftrightarrow A_{(s)} + xH_2O_{(g)} \tag{6.6}$$

where the subscripts (s) and (g) refer to solid and gas phases, respectively. This reversible reaction is analogous to the chemical reaction in Eq. (6.5). The desorption process from left to right is endothermic and requires the input of energy (typically heat). This energy is recovered by exposing the adsorbate to water vapor, as represented by the exothermic reaction from right to left.

Material that is chosen for latent heat storage should satisfy the following criteria.

- high adsorption heat
- high capacity for water adsorption
- reasonable adsorption temperature

- fast adsorption/desorption reactions

While a number of materials have been used for adsorption energy storage, and there is much active research in the area, two traditional materials, zeolites and silica gel have attracted considerable interest and are reviewed briefly below.

6.5.2.1 Zeolites

Zeolites are aluminosilicates with an open three-dimensional framework consisting of corner-sharing oxide tetrahedra. They have the general formula

$$M^{n+}_{1/n}(AlO_2)^-(SiO_2)x \cdot yH_2O, \tag{6.7}$$

where M is either a metal ion or H^+. Zeolites occur naturally and can also be produced synthetically. The structure of a typical zeolite (mordenite) is illustrated in Fig. 6.12, which shows the interconnected SiO_4 tetrahedra.

Zeolites, both those that are naturally occurring and those that are synthesized, are used extensively for their adsorptive properties. The pores formed between the oxide tetrahedra are of a size that is compatible with the size of atomic and molecular species. The pore size varies from about 0.1 nm to about 1.0 nm. This can be controlled by adjusting the composition of a synthetic zeolite or choosing a natural zeolite with suitable properties.

Results of a modelling investigation of the adsorption energy of the zeolite 13X – water system are illustrated in Fig. 6.13. The 13X zeolite has a nominal composition $Na_2O{\cdot}Al_2O_3{\cdot}2.8(SiO_2){\cdot}(6\text{–}7)H_2O$ and a mean pore size of 1.0 nm. The study shows a

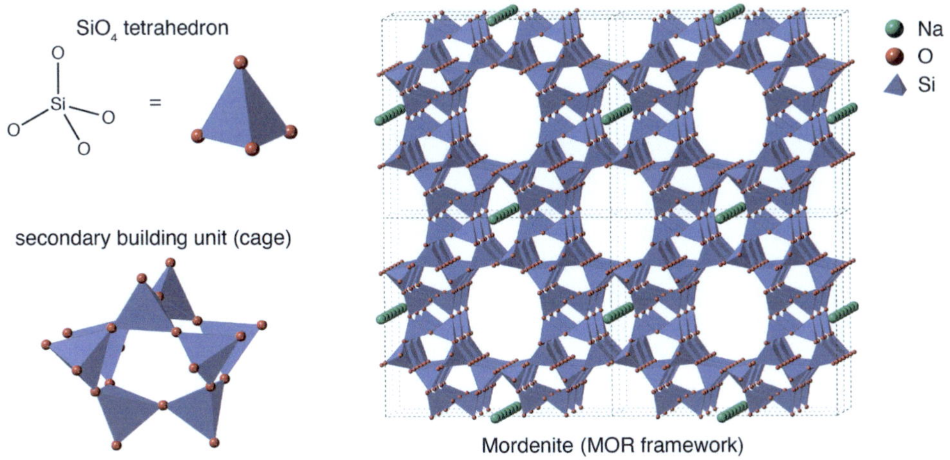

Fig. 6.12 The structure of the zeolite mordenite (AlSiO·7HO) showing the component SiO_4 tetrahedra. Mordenite is named after the community of Morden, Nova Scotia, Canada, where it was first identified in 1864. Coudert (2020) CC BY 4.0. https://creativecommons.org/licenses/by/4.0/deed.en

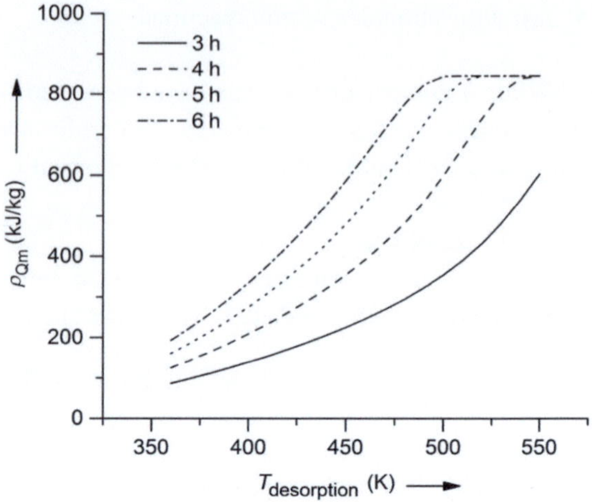

Fig. 6.13 Modelling results for energy density of the zeolite (13X)-water system showing adsorption energy as a function of desorption temperature for different desorption times. Figure 2 from Kohler and Müller (2017) Copyright (2017) The Authors CC BY 3.0. https://creativecommons.org/licenses/by/3.0/

maximum energy storage capacity of 844 kJ/kg. Using a bulk density of 13X of 750 kg/m^3 (Xi'an Lvneng Purification Technology Co., Ltd. 2024), gives a storage capacity of 176 kWh/m^3. This modelling result is consistent with experimental investigations of 13X (e.g., Dicaire and Tezel 2013).

6.5.2.2 Silica Gel

Silica gel is a synthetic amorphous form of silicon dioxide (SiO_2) containing nanometer sized voids and pores. These voids and pores, which average around 2.4 nm in size, are of suitable size for the adsorption of water and various organic molecules. Figure 6.14 shows the surface of silica gel and the bonding of water molecules to the surface atoms.

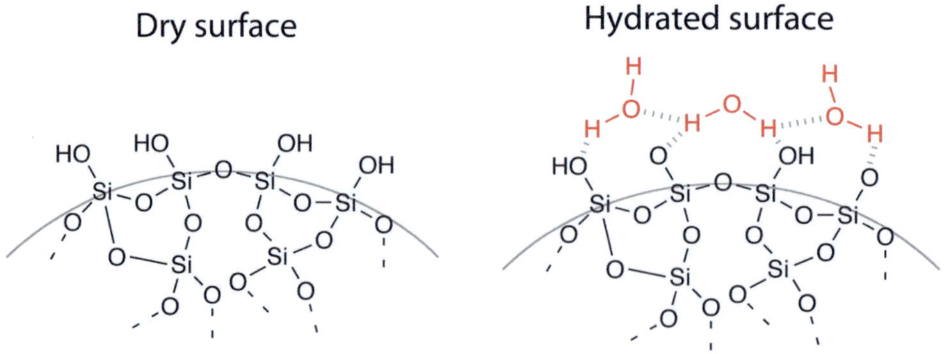

Fig. 6.14 Surface of silica gel without (left) and with (right) adsorbed water molecules. Mons (2007) CC BY-SA 3.0. https://creativecommons.org/licenses/by-sa/3.0/deed.en (labels translated to English)

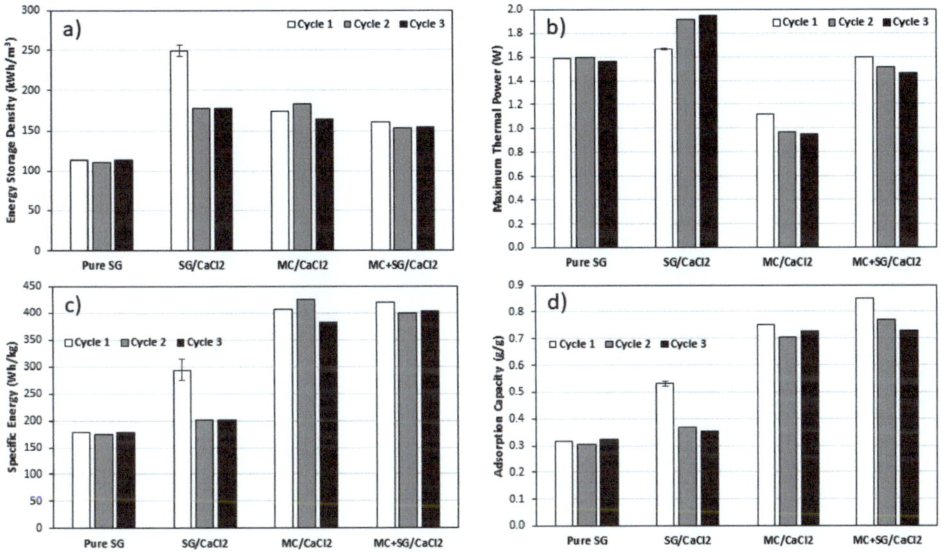

Fig. 6.15 **a** Energy storage density, **b** maximum thermal power, **c** specific energy, **d** water–vapor uptake capacity for three hydration/dehydration cycles performed at adsorption inlet relative humidity of 50%. Samples are pure silica gel, methyl cellulose and mixtures including $CaCl_2$. Figure 8 from Shervani et al. (2024) With permission of Springer. CC BY 4.0. https://creativecommons.org/licenses/by/4.0/

The results of an investigation of the adsorption energy storage capacity of silica gel and mixtures including silica gel are illustrated in Fig. 6.15. The exact energy storage capacity depends on factors such as particle size, relative humidity and adsorption/desorption temperature. Figure 6.16 illustrates the effects of relative humidity on the energy storage capacity of silica gel-based storage media. Typical maximum storage capacity found was found to be of the order of 200 kWh/m^3.

The above results for adsorption energy storage using zeolites and silica gel can be compared with the energy storage capacity of other systems. In Table 6.3. the energy storage capacity in kWh/m^3 (commonly known as the volumetric energy density) is given for some common energy storage technologies. Results for adsorption energy storage capacity using zeolites and silica gel are somewhat less than modern battery technologies (e.g., lithium-ion) but are better than traditional battery chemistries (e.g., lead-acid). Most notably, however, they are far superior to mechanical methods, such as pumped hydroelectric storage and compressed air storage, as described in Chaps. 1 and 2, respectively.

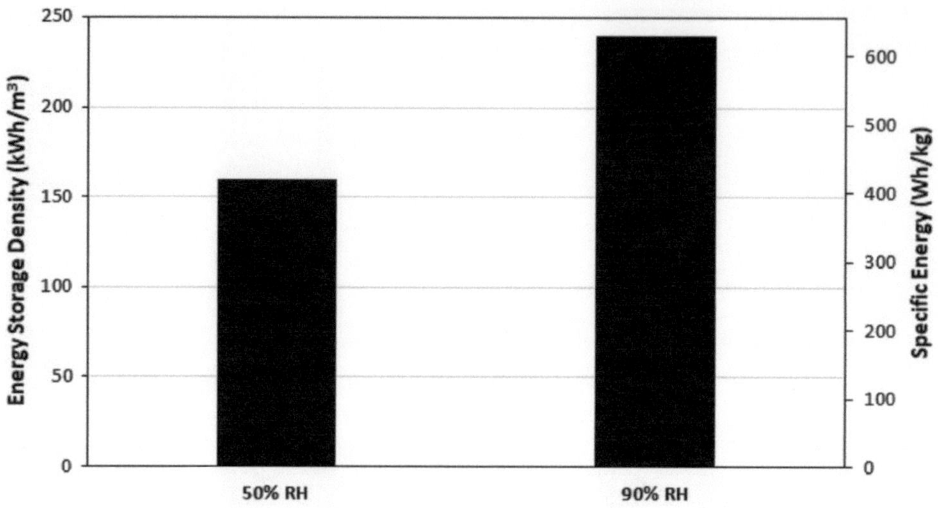

Fig. 6.16 The energy storage density and the specific energy for MC/SG/$CaCl_2$ at 50 and 90% inlet relative humidity. Data are averages of first three cycles as given in Fig. 6.15. Figure 10 from Shervani et al. (2024) With permission of Springer. CC BY 4.0. https://creativecommons.org/licenses/by/4.0/

Table 6.3 Typical energy content per m^3 (volumetric energy density) for different energy storage methods

Storage method	kWh/m^3
Zeolite 13X	176
Silica gel	200
Pb-acid battery	85
Li-ion battery	680
100 m head hydroelectric	0.28
10 MPa compressed air	13

References

Astronaut (2014) Height of the Burj Khalifa compared to some other well known tall structures. https://commons.wikimedia.org/wiki/File:Burj_Khalifa_Height.svg

Banaei A, Zanj A (2021) A review on the challenges of using zeolite 13X as heat storage systems for the residential sector Energies 14:8062. https://doi.org/10.3390/en14238062

Cabeza LF, Castell A, Barreneche C, de Gracia A, Fernández AI (2011) Materials used as PCM in thermal energy storage in buildings: a review. Renew Sustain Energy Rev 15:1675–1695. https://doi.org/10.1016/j.rser.2010.11.018

Costa MC, Sardo M, Rolemberg MP, Coutinho JAP, Meirelles AJA, Ribeiro-Claro P, Krähenbühl MA (2009) The solid–liquid phase diagrams of binary mixtures of consecutive, even saturated fatty acids. Chem Phys Lipids 160:85–97. https://doi.org/10.1016/j.chemphyslip.2009.05.004

Coudert F-X (2020) Zeolite structure as an assembly of tetrahedra. https://commons.wikimedia.org/wiki/File:Zeolite_structure_as_an_assembly_of_tetrahedra.png

Ding YL, Tong LG, Zhang PK, Li YL, Radcliffe J, Wang L (2016) Liquid air energy storage, chap 9. In: Letcher TM (ed) Storing energy—with special reference to renewable energy sources. Elsevier, Amsterdam, pp 167–181

Dicaire D, Tezel FH (2013) Use of adsorbents for thermal energy storage of solar or excess heat: improvement of energy density. Int J Energy Res 37:1059–1068. https://doi.org/10.1002/er.2913

Dumpygrimbo (2024) Diagram describing flow of air and heat through a cryogenic energy storage system. https://commons.wikimedia.org/wiki/File:Cryogenic_Thermal_Storage_Diagram.png

Kerskes H (2016) Thermochemical energy storage, chap 17. In: Letcher TM (ed) Storing energy—with special reference to renewable energy sources. Elsevier, Amsterdam, pp 345−372. https://doi.org/10.1016/B978-0-12-803440-8.00017-8

Kohler T, Müller K (2017) Influence of different adsorbates on the efficiency of thermochemical energy storage. Energy Sci Eng 5:21–29. https://doi.org/10.1002/ese3.148

Mons S (2007) Surface structure of silica gel, dry and hydrated. https://commons.wikimedia.org/wiki/File:Schematic_silica_gel_surface.png

Naresh R, Parameshwaran R, Ram VV, Srinivas PV (2022) Study on thermal energy storage properties of bio-based n-dodecanoic acid/fly ash as a novel shape-stabilized phase change material. Case Stud Thermal Eng 30:101707. https://doi.org/10.1016/j.csite.2021.101707

Noël JA, Kahwaji S, Desgrosseilliers L, Groulx D, White MA (2022) Phase change materials, chap 22. In: Letcher TM (ed) Storing energy—with special reference to renewable energy sources, 2nd edn. Elsevier, Amsterdam, pp 503−535. https://doi.org/10.1016/B978-0-12-824510-1.00005-2

Ristić A (2022) Sorption material developments for TES applications, chap 27. In: Hauer A (ed) Advances in energy storage—latest developments from R&D to the market. Wiley, Hoboken, NJ, pp 631−653

Shervani S, Strong C, Tezel FH (2024) Simultaneous impregnation and microencapsulation of $CaCl_2$ using silica gel and methyl cellulose for thermal energy storage applications. Sci Rep 14:7183. https://doi.org/10.1038/s41598-023-50672-6

Xi'an Lvneng Purification Technology Co., Ltd (2024) Zeolite 13X molecular sieve desiccant pellet. https://www.molecularsieveadsorbent.com/sale-38408120-zeolite-13x-molecular-sieve-desiccant-pellet-bulk-density-0-75g-ml.html

Zbair M, Bennici S (2021) Survey summary on salts hydrates and composites used in thermochemical sorption heat storage: a review. Energies 14:3105. https://doi.org/10.3390/en14113105

MIX
Papier aus verantwortungsvollen Quellen
Paper from responsible sources
FSC® C105338

If you have any concerns about our products,
you can contact us on
ProductSafety@springernature.com

In case Publisher is established outside the EU,
the EU authorized representative is:
Springer Nature Customer Service Center GmbH
Europaplatz 3, 69115 Heidelberg, Germany

Printed by Libri Plureos GmbH
in Hamburg, Germany